CONCEPT AND EXPLANATION IN BIOLOGY

Edited with Introductions by
Irwin Savodnik, M.D., Ph.D.

distributed by
DABOR SCIENCE PUBLICATIONS
Oceanside, New York 11572

Introduction © 1977 by
Irwin Savodnik

Library of Congress Cataloging in Publication Data

Main entry under title:

Concept and explanation in biology.

  A collection of selections from various sources.
  Bibliography: p.
  1. Biology--Addresses, essays, lectures.
I. Savodnik, Irwin. QH311.C64 1977 574'.08
77-11000
ISBN 0-89561-032-9

A Beau Baker Book

Printed in the United States of America

ACKNOWLEDGEMENTS

The editor is grateful to the following publishers for their permission to reprint selections from the works listed below.

Cambridge University Press.
  Braithwaite, R.B., *Scientific Explanation*, Chapter X.
  Dobbs, H.A.C., *British Journal For The Philosophy of Science*, VII, pp. 140-150.
  Kapp, R.O., *British Journal For The Philosophy of Science*, V, pp. 91-103.
  Russell, E.S., *British Journal For The Philosophy of Science*, I, pp. 108-116.
  Schrodinger, E., *What is Life?*, Chapters VI & VII.

Harcourt Brace Jovanovich, Inc.
  Nagel, Ernest, *The Structure of Science*, 1961, pp. 428-446.

Harper & Row Publishers, Inc.
  Bertalanffy, Ludwig Von, *Modern Theories of Development*, 1933, pp. 1-27 in 1962 edition.
  Bertalanffy, Ludwig Von, *Problems of Life*, 1952, pp. 9-22.

Holt, Rinehart and Winston.
  Bergson, Henri, *Creative Evolution*, translated by Arthur Mitchell, pp. 97-108.
  Morgan, Thomas, *The Mechanism of Mendelian Heredity*.

The Quaterly Review of Biology.
  Woodger, J.H., V, pp. 1-22.

Routledge & Kegan Paul, Ltd.
  Nagel, Ernest, *The Structure of Science*, pp. 428-446 (British rights).

University of North Carolina Press.
  Sinnott, Edmund, *Cell and Psyche*, 1950, pp. 15-42.

CONTENTS

To My Parents

PREFACE

As the philosophy and history of biology grows, so too, does interest in the origins of the problems which constitute it as an area of philosophical concern. With this interest in mind, I have sought to select segments from those sources which not only constitute clear statements of the different points of view but also serve as the original expression of those statements. While this has not been possible with every selection, the overall perspective of the book certainly embodies this attitude. Hence, a good number of excellent articles of more contemporary vintage have been excluded from this volume. Hopefully, the bibliography will provide the reader with sources for further reading. While some authors included here may have had the first word, the least I can do is point the way toward the last, for the time being at any rate.

The conceptual issues which are expressed here serve as examples of the manner in which the philosophy of biology has approached the problems of understanding the nature of living things. Thus, the debate between mechanism and vitalism is an example of an almost timeless controversy about the character of the organism. While molecular biologists may think that they have put the problem to rest, others who maintain a certain positive attitude toward the autonomy of biological organization still insist that the issue is very much alive. In any case, such issues are never as simple as they first appear and a careful reading of the various points of view reveals subtleties of argument not previously seen before.

These arguments about the nature of life contain a larger vision of philosophical concern. There is, of course, the cosmic perspective of a Bergson or a Whitehead and the view of the former is offered in this book. There is also, however, the issue of the difference between human beings and non-living things with regard to the issue of autonomy, self-direction and intentional behavior. Behaviorism is to the metaphysical view of man what mechanism is to the metaphysical view of life. Similarly with intentionalism and vitalism. Both sets of issues seem to point to a more inclusive division of philosophical postures concerned with an ultimate view of nature with particular respect to the role that animated entities play within such a context.

Footnotes have been preserved as they appeared in the original selections.

This volume has been stimulated by a number of different individuals. Richard Martin directed my doctoral dissertation in the philosophy of biology and served as a model for thoughtful perseverance. Kenneth Stern has served as a sounding board for endless hours—far more profitable ones for me than him. My wife, Marlene Savodnik, has persisted with me in this task for longer than either of us like to think.

Cynthia Freeland compiled the bibliography and provided invaluable editorial assistance. Mary Lou Borgen typed the entire manuscript several times and assisted in a multitude of ways. I would also like to acknowledge the support of the Department of Psychiatry at the University of Pittsburgh for providing me with the necessary time and assistance in bringing this work to fruition.

INTRODUCTION

Within recent decades biology has played an increasingly significant role in the life of man. There has been an acceleration of discovery in the biological arena which is unequaled in all its history. From the point of view of the philosopher, the historian, indeed the biologist himself, this increasing rate of development has tended to blur the fundamental issues of biology as a science of life. It is from this perspective that this volume is presented. The major goal of this book is to elucidate some of the basic issues of biology through a presentation of those writings which, in the opinion of the editor, most clearly illustrate the seminal issues of the science. There are three major paths whereby this clarification may be initiated and concomitantly, these volumes are subdivided along these same lines. There are two parts to this volume, each of which throws light on fundamental areas of biological concern from a different direction. In a second volume a series of related but distinct issues are addressed which complement the works in this volume. Each book, however, stands on its own. The two volumes together constitute a comprehensive approach to the fundamental questions of biology.

In Part I all of the authors address themselves in one way or another to the question *What is life*? Peculiarly, this question is really one of philosophical or conceptual concern rather than purely biological. The reason for this is that the problem is one which depends not just on the acquisition of additional data, but instead on the effort to find an incisive road into a maze of data and make sense out of it. In Aristotle's work *De Anima*, for instance, there is no attempt to prove a point on the basis of an empirical study. Rather, he considers the soul in a way that does not ignore his profound knowledge of living things but that is oriented at bringing about some order amongst all that he has learned about living things, particularly human things. The soul, then, is a power which is possessed by all living things and something that sets the living world apart from the non-living world. It is, briefly, the principle of life. Similarly, Driesch proceeds to indicate on a conceptual rather than an empirical plane that what he considers mechanical explanation is incapable of accounting for the diverse phenomena of living things, but that there must exist such a thing known as an entelechy, which is conceived of as very similar to the soul which Aristotle postulates. He certainly relies on his voluminous research as a biologist; but he does not attempt to prove his point empirically. He recognizes that the problem is not purely empirical but rather one requiring philosophical analysis.[1]

The major themes of the first part of this book are those of vitalism, mechanism, and organismic biology. These will be discussed in the introduction to that section of the book. It should be indicated though, that each of these perspectives represents a series of attempts to answer that fundamental question about the nature of living things. It is fascinating to see that the conflict between vitalism and mechanism (organismic biology is a relative newcomer to the scene) is a seemingly never ending one. Descartes and La Mettrie present two of the most famous and provocative arguments for the mechanistic view of life that one can read. However, the fact that the problem is not yet solved—indeed, it often rages in a paroxysmal manner, often waning for many years but never dying out entirely—is not an indication of its worthlessness. On the contrary, as the battle rages, newer and more sophisticated techniques of analysis and argument are introduced, along with the increasingly complex data that are uncovered in biological exploration. The progress of this long historical debate reveals the height to which the intellect of man can be elevated in an attempt to deal adequately with a tantalizing and basic problem of biology. The related discussions in this section of the book are of interest to the philosopher because they help to clarify certain issues in the philosophy of biology. The biologist will, it is hoped, find particular appeal in this portion of the book since he is rarely afforded much time for such luxurious considerations, even when these thoughts underlie a great deal of what he does in the laboratory. In a sense, then, a common arena is presented where the philosopher and biologist can meet to discuss issues of interest to them both.

Part II of the book is concerned with biological methodology and explanation. The four selections point out the central concerns of philosophers and biologists when they consider the theoretical underpinnings of biological procedure and explanation. Inevitably, the question arises as to the relationship biology has or ought to have with physics. This question is present either implicitly or explicitly in these selections. When it is investigated more closely, it is seen that there are several questions which are really being asked. We may list them as follows:

1) Can physics serve as a model for biology with regard to the entities studied?
2) Is biological methodology merely a branch of physical methodology?
3) Can the categories of physics suffice for all cases of biological explanation? Or,
4) Is there an autonomous biological mode of explanation?

This list certainly does not exhaust all the questions related to the problem of the interface of biology and physics. It serves, however, to indicate some of the major areas of consideration which philosophers have established when they have dealt with the problem. The first question, for instance, raises many subsidiary problems whose investigation helps to throw light on the nature of biological investigation. In the first place, one wonders just what kind of a science biology really is. Are biologists just like physicists except for the fact that they are dealing with different subject matters? If so, then the answer to the first question should be a definite yes. However, the answer to the question is anything but simple. In reading about biology, two interesting features become evident. First, biology is not as reflective a science as is physics. That is, it does not question its own foundations as rigorously as does physics. Second, biology is not often the subject of the philosopher's concern. More often than not, physics serves as the paradigm case of an empirical science for philosophers. When most philosophers think of the philosophy of science, they generally think of the philosophy of physics. It is quite true that there are many people interested in the behavioral sciences from a philosophical standpoint, but it is hard to deny that these disciplines hold a lesser position with respect to the amount of time and energy philosophers devote to the problems of science. There are many different explanations for the high position which physics seems to have attained for itself in the hierarchy of the sciences. A major factor is the fact that more than any other science, physics has fruitfully incorporated mathematics into itself—to the extent that, were mathematics removed from the corpus of physics, the science would cease to exist as we know it today. Indeed, other sciences seem to use physics as a model science insofar as it uses mathematics to the extent that it does. The mathematization of physics encourages the formalization of the science, and it is this latter feature which renders physics such a popular science for philosophical analysis.

With formalization, one is able to clarify the status of the various kinds of statements that are found within the framework of the science. Postulates, hypotheses, empirical statements, and laws are all clearly distinguishable from one another. Therefore, the architecture of the science becomes transparent and the science becomes amenable to analysis. Physics has developed to the point that many of its branches can be presented as deductive models bearing the necessity of logical argument.

Biology, for whatever reasons may be offered, has not enjoyed the degree of formalization of physics. Given this condition, we can see why biology has not achieved the same status as physics in the eyes of the philosopher and, indeed, in the eyes of the biologist. For if the latter were to think of his science as being as highly developed as physics then the lack of reflection on the part of the biologist mentioned earlier would not be apparent. It is certainly the case that mathematics is a newcomer to the science of biology. Classical biology, which extends to the mid-twentieth century, is to the new biology what classical synthetic geometry was to the new 17th century science of analytic geometry. The translation of statements about biological events into mathematical statements is an enormous step for the science of biology. While it may be said that it is a necessary condition for biology to reach the same level as physics in the eyes of the philosopher, it is certainly not sufficient.

It is often pointed out that biology has not yet found its Newton. Certainly, it has not yet found its Einstein. The point here is that until recently the science had not yet undergone the type of revolution which physics has experienced at least twice already. Without such upheaval within the confines of the biological world it is hard to imagine that biology will ever reach the same plane as physics. However, the last few years in biological thought have provided for modern man the profound possibility that a true biological revolution is actually occurring. With the discovery of the structure of the genetic material of all living things and the consequent development of this discovery, it is quite conceivable—indeed it is probably the case—that biology has undergone a fundamental metamorphosis which has altered its view of living reality from ground up. As with all revolutions, however, it is difficult to assess where biology is going and whether the recent developments which have been uncovered will reap a true revolutionary harvest.

It seems, then, that biology is in something of an amphibious position with one foot in the realm of tradition and the other in the whirlpool of revolution. Certainly, it is at a fascinating juncture in its history. Because of the rapid changes which have occurred in the science so recently, the question as to whether or not physics should serve as a model for biology is heightened both in importance and difficulty. For if we are to make any sense out of the science, then we ought to know what sort of direction it will take in the future, and the question of its relation to physics is explicitly aimed at finding out what that direction will be.

Another point which tends to relegate biology to a secondary position among the sciences is that it utilizes physical principles in the explanation of living phenomena. The question must arise, then, as to whether or not biology is merely a branch of physics. At first glance this question might seem rather easy to answer regardless of the solution offered. It is a peculiar characteristic of this problem that both sides seem to think that they are not only right but also that their answers are so obvious as to be almost self-evident. However, as the selections by Bertalanffy, Woodger, and Nagel indicate, there is no simple solution to the problem. The initial temptation of one who subscribes to the simple reductionist point of view is to state that the reason biology has not yet been entirely incorporated into physics and still maintains a semblance of its own identity is that the procedures of biology are still so fraught with inaccuracies that the precise methods of physics can be only of limited help. It remains for the biologist to introduce the accuracy of physics into his empirical procedures before the clarity of physical explanation is completely attained. This attitude is very much the same as that taken towards the social sciences by many people. They, too, feel that the introduction of better techniques into the social sciences is all that lies in the way of "physicalizing" these areas of human investigation.

There are those, however, who see the problem quite differently. Biology, for these individuals—Bertalanffy amongst them—is not to be thought of merely as a branch of physics because the nature of biological organization is unlike any type of organization studied in physics. This leads us into a consideration of the third question mentioned above. Many individuals assert that while a great deal of the principles and procedures of physics may be incorporated into biology, it is not the case that biology can be reduced to physics for the simple reason that physics does not have an entire set of conceptual tools necessary for a thorough accounting of all biological events. Certainly, the concepts of physics as they now exist do not suffice for biological purposes. There is no concept in physics which is substitutable for that of "inducer" in biology. The problem then arises as to whether or not it is possible to translate such a term as "inducer" from the vocabulary of biology to that of physics and then provide a useful explanation of a biological phenomenon in purely physical terms. This problem becomes highly involved as we begin to distinguish the various types of meanings which terms like "physical explanation" and "mechanical explanation" have in varying contexts. One of the best examples of the relationship between causal explanation as it is employed in physics and teleological explanation as it is used in biology and psychology is to be found in the article by R. B. Braithwaite in this volume. The problem of the translatability of the two modes of explanation is discussed. In addition, Braithwaite considers the question of whether or not there is an autonomous biological mode of explanation. The problem of an autonomous mode of explanation in biology can be viewed in different ways, and it is not unprofitable to look at it in the way Braithwaite does, that is to opt for the best of both possible worlds. While the task of translating biological statements into physical ones might be a theoretical possibility, it might not be a very desirable thing to do since the principles of the science may find a clearer elucidation in so-called biological terms rather than physical ones. The full argument, however, is left to the consideration of the reader.

The question of explanations in biology versus physics is still an open one, and the selections presented in this volume are considered to be some of the most precise and historically significant statements about this problem. What cannot be presented in this volume is the nature of explanation in the physical sciences. If a counterattack is to be launched by the "physicalists," it would seem to be along lines which suggest that physical explanation is capable of far greater things than thought of by those who defend the autonomy of biological explanation. Here a point which has been made before ought to be emphasized. It is one thing to use physical techniques and modes of analysis in biology and quite another to usurp what is properly within the domain of the biologist. The great danger of the latter is that biology will be denied its chance to develop its own potential for comprehending the living world to its fullest extent. If the tools of the more advanced sciences are introduced too early, then the conceptual power of the biologist as well as his ability to develop novel modes of investigation may be truncated. A good example of advanced techniques being imposed on a science prematurely is to be found in the development of geometry. The introduction of analytic techniques to geometry is certainly to be viewed as significant with respect to the development of calculus and Newtonian physics.

With the new use of algebraic tools in the statement of geometrical relationships the dynamic structure of physical events could be more clearly represented. Certainly, these advances cannot be ignored. However, what is often forgotten is the price that was paid by traditional geometry in its almost total transition to analytic geometry.

What was not realized at the time of this transition is that traditional geometry had a great deal more development to undergo within its own mode of conceptualization. It had not exhausted its own possibilities, and the imposition of a new set of analytical techniques on the science retarded the growth of geometry by causing a de-emphasis of the techniques traditionally used. It was not until the 19th century that the developments of non-Euclidean geometry indicated the extent to which the potentialities of classical geometry could be realized. With the emergence of non-Euclidean geometry, the traditional modes of geometrical analysis proved to be powerful and enlightening. The advance of analytic geometry had proved to be a mixed blessing. For with this move forward there was a concomitant step in the reverse direction.

In the case of biology, the premature imposition of purely physical concepts on the domain of biology and the substitution of strict physical explanation for what is now regarded as biological explanation may result in a situation similar to that of geometry. If biology is not given the opportunity to develop to the fullest its own conceptual apparatus, then it may severely be limiting its own potentialities. In addition, the full depth of the various problems which are investigated in biology might never be plumbed because the symbolic apparatus utilized in the imposed mode of explanation is not fitted well enough to the job. Thus, there is the very real possibility that the profundity of the biological sphere may be sacrificed to the accuracy of physical explanation. This accuracy becomes pernicious because it causes biology to abandon what is truly its own. For that to happen to any science is most certainly a tragedy of the first order.

In the companion volume to this work[1] there is a series of articles that are central to the historical development of biology. The attitude taken here is that contemporary ideas cannot be fully appreciated without some knowledge of their origins. The true profundity of the Watson-Crick model can only be recognized after the long historical emergence of the science of genetics is realized. To read those last few chapters of *The Double Helix* is in essence to read the final chapters in a much longer story played out in a period of time spanning two centuries. In addition, what seems to be a final chapter in reality turns out to be the opening pages of a book which is still being written. The drama of the development of biological ideas is in a certain sense necessary to the understanding of these ideas.

In viewing the historical developments in biology, consideration must be given to the relation biology enjoys with other disciplines. Of particular importance is its relation with medicine. It is in this second volume that the interaction of the two can be seen at least partially. The problem of how the two areas are interrelated raises some interesting questions concerning the role which biology ought to play in our lives. There are those involved in medical education, for instance, who feel that biology is a mere handmaiden to medicine. What is not directly relevant to the study or practice of medicine, is, from this point of view, essentially unimportant biology. Concrete manifestations of this attitude are apparent when one examines the criteria employed in determining the distribution of funds for government grants to biological research. While the first reaction to this view might be one of proper outrage at the obvious philistinism of its advocates, it is important to keep in mind that human well-being and physical health will generally take priority in the minds of most men to research which has no apparent application to the betterment of man, other than the possible advancement of his knowledge about what makes things alive rather than not alive. However, there is a better way to look at the problem, and that is to point out that the fruits of biology are very difficult to predict. Who could state with any accuracy the future implications of the work of Mendel, Darwin, and Pasteur? Yet, these three men have had an overwhelming influence on the science of microbiology. In addition, there is a sort of emergent quality about the growth of biological knowledge. That is, given the work of Darwin on natural selection, and that of Mendel on genetics, and finally that of Pasteur on microbes, the effect of their work taken together on the development of microbiology is not readily predictable from an examination of their separate accomplishments. After all, who would think that the scale of Darwin's evolutionary conceptualization of living things could be reduced to that of microscopic organisms whose generations are measured in hours rather than in decades? Furthermore, consider the difficulty in predicting that the work of Mendel would some day awaken men to look for a substance indigenous to every living thing which would enable them to predict the probabilities of microorganisms having various propensities towards the fermentation of various sugars. Certainly, these developments, which have occurred with an accelerating frequency, indicate that any attempt to narrow biological research or even thought to that realm which has only direct applications to the field of medicine is self-defeating on its own grounds, since the predictability of the outcome of various biological discoveries is extraordinarily limited. (Witness also the example of the development of non-Euclidean geometry which found direct application in the cosmologies of this century. In no way could such an application have been predicted.)

On another level, however, there is a severe contraindication to the narrowing of biological thought on the basis of its applicability to any field whatever. To act in such a way would simply be contrary to the manner in which the human spirit is accustomed to act. In order for the greatest advancements in learning to take place, it is necessary that the human mind function in a manner unbridled by any considerations other than those which directly relate to the problem at hand. To ask for anything more is to confuse human knowledge with something far less and perverse. Just as music cannot be an instrument solely for the use of the state, so biology cannot be a tool strictly in the service of medicine. Indeed, to leave it unhampered by any such restrictions is in the best interests of medicine. Only in the unlimited atmosphere of human curiosity and ceaseless investigation will the problems of modern medicine be answered. Here the progress of knowledge parallels that of life itself. Living things are not terribly predictable in their paths of development. The greatest achievement of the living sphere is found in the free domain of the evolutionary process. Witness the emergence of man—for better or worse.

NOTES

[1] A philosopher such as W. V. O. Quine would have objections to the distinction between "conceptual" and "empirical" in this context, but for the purposes of this presentation such objections are not crucial. In an extended argument, however, Quine's attack on such distinctions would certainly have to be acknowledged. See Quine, "Two Dogmas of Empiricism," in *From a Logical Point of View*, New York: Harper & Row, 1961.

[2] *The Origin of Structure in Biology*, Dabor Science Publications, 1977.

Part I

INTRODUCTION

While all biologists address themselves to the question, What is life?, relatively few have sought to offer an explicit and direct answer to that question. The following selections are chosen from that group of thinkers whose aim it is to seek an understanding of the unique nature of the process of life. While at first there may seem to be a large variety of answers, in fact there are relatively few types of replies which have emerged. Many of the selections represent variations on the same theme, and it is often a study of these variations that yields the more subtle elements of a particular position. I will attempt to outline the basic modes of approaching the question of the nature of life in order that the selections taken together can be seen in a unified way.

The major problem to which all these thinkers address themselves is that of understanding what it is about living things that differentiates them from non-living things. After all, living things are made up of the very same material as inanimate objects, and yet they are somehow different. The more one reflects about this fact, the more intriguing it becomes. Aristotle sees the problem quite clearly and asks a question that may be phrased as follows: Why are things alive rather than not alive? There must be, says Aristotle, something that renders a living thing alive rather than not alive, and that something is not immediately evident to the unaided eye. Indeed, visual perception will not lead to the soul in the sense that anatomical dissection will lead to the organs of the body. For the soul is not a special part of the body. Rather, the soul is related to the body in a way that can only be understood in the larger framework of Aristotle's philosophy. All things are composed of matter and form, and insofar as they have matter they have potentiality. Similarly, the form of a body corresponds with the actuality of that body. In the case of a living thing, the soul is the form of the body. It is the first actuality (entelechy) of a body having life in potency, i.e., a body

1

organized naturally so as to be able to perform various life functions. The soul, then, for Aristotle is the principle of life, and an explication of its nature is necessary for an understanding of living phenomena. All living things have souls. That is why they are alive. There is a hierarchy of souls, the soul of man occupying a relatively high position on the chain of being which Aristotle constructs. Lower souls of plants are capable only of fulfilling the nutritive faculty; animals' souls have in addition the faculty of appetition. Higher souls, namely, the souls of man, are capable of thought. A body, then, in itself is not alive, but rather, has life in that it is informed by a soul.

Aristotle is perhaps the first vitalist. Vitalism will be discussed in more detail later, but it suffices to say at the moment that its distinguishing characteristic is that a vital principal which is incorporeal is postulated as the factor that makes a thing alive. Descartes holds a sharply opposite viewpoint from that of Aristotle. Living bodies are no more than mechanisms which are in no need of souls to explain their living nature. This is easily seen in the case of animals, which are composed strictly of corporeal substance. If they were in part made up of spiritual substance, then they would manifest the essential property of the latter, namely thought. But this is not the case with animals. Some beasts do indeed display behavior which might be construed as thoughtful, but there is nothing in their corporeal nature. There is no need to postulate a soul. Human beings, however, are composed of both corporeal and spiritual substance. They have both body and soul. However, their life processes are not at all dependent upon the intervention of the soul. The beating of the heart, for instance, requires no spiritual aid. It is a purely mechanical process as Harvey had shown. The human body is a machine or an automaton. It is only because men are capable of rational thought that we can be sure that they have souls. Of course, this complete divorce of the realms of thought and life-processes creates an acute problem for Descartes in explaining, for example, how a thought or wish can lead someone to move in a certain way. The body and the soul *do* interact, he was eventually forced to say, and this problematic interaction takes place in the midline structure of the nervous system called the pineal gland. The mind does not move man. Instead, it is conceived as directing movements, the power of which resides solely in the body. This power of movement, then, has nothing to do with the presence or absence of any special vital principle. The distinguishing characteristic between living and non-living things, for Descartes, is simply the state of the machine and nothing else. The dead body is analogous to a watch unwound; the vital principle simply does not enter the picture.

The mechanism which Descartes presents is but an ancestral hint at the results which the 18th-century physician La Mettrie would arrive at. Having indicated the difference between the Aristotelian conception of the soul and that of Descartes, it is a natural step to turn to the polemical writings of La Mettrie, who develops the concept of mechanism as fully as possible in his day. The very title gives away the meaning of the work, and there is little to add. Together with Descartes, La Mettrie provides a powerful anti-Aristotelian and anti-vitalist argument in his writings. La Mettrie writes with a biting wit not often found in scientific or philosophical reflections. The following is but one example among many from the selection in this text: "Let us observe the ape, the beaver, the elephant, etc., in their operations. If it is clear that these activities cannot be performed without intelligence, why refuse intelligence to these animals? And if you grant them a soul, you are lost, you fanatics!" At the very least, then, La Mettrie provides us with an articulate and lively presentation of the mechanistic position.

With the mechanist point of view firmly presented, the writing of Hans Driesch serves as the finest example of the opposing position, i.e. vitalism. Driesch worked as an embryologist, his most renowned work being that done on the sea-urchin embryo. It was on the basis of his own experiments that he arrived at his formulation of vitalism. The sea-urchin egg begins its embryonic itinerary by first dividing into two cells, then four, then eight, and so on, until more complex germ layers are derived. In one series of experiments, Driesch took a sea-urchin egg just as it was dividing into the two-cell stage and separated the two cells. He then observed that both cells developed into whole sea urchins which were a bit smaller than ones that had not been subject to this treatment but were otherwise entirely normal. It was reasoned by Driesch that if the sea-urchin embryo were divided into two, there should be the concomitant development of two half sea-urchins rather than one whole organism. But this was not the case, and Driesch responded to his initial reaction by postulating a guiding force which led each half to full development. This postulate implied that the laws of physics and chemistry were incapable of explaining the phenomenon which Driesch had observed. There was need for an extra-material entity which he called an *entelechy*, after Aristotle, to explain what had happened. The organism, then, for Driesch, can not be a machine. The division of machines into halves results in half-machines, not wholes. There must be an entelechy, a vital force which maintains the structural and developmental integrity of the organism.

Today, vitalism is not regarded with much respect by most biologists and philosophers. The postulation of such a non-observable entity in a science which purports to be strictly empirical is thought to be antagonistic to the very foundations of that science. But there is more to Driesch than the mere formulation of the idea of a vital force. Driesch recognized some of the most profound aspects of living systems that were not fully appreciated by his predecessors. In particular, he understood that the dimension of time was a central element in the existence of the organism. In his logic the concept of *becoming* is an irreducible one. In taking this position Driesch departs from other logics, which held that it is quite possible to do without such concepts as becoming and change. For Driesch, an understanding of living things must take as its basis the processive nature of the organism. Mere structural analysis cannot account for such biological phenomena as adaptation and regeneration. There must be a transcendent power which directs the organism in a direction most suited for its own survival.

The full metaphysical implications of Driesch's thought are worked out in Bergson's *Creative Evolution*. Today, Bergson is looked upon as being of interest largely from an historical point of view. This is unfortunate in that he has a great number of insights which, while couched in difficult and often obscure language, are nevertheless fruitful when thought through to their ultimate conclusions. At one time during his life, Bergson's philosophy was widely studied by all those interested in the great contemporary movements in Western thought. Bergson rejects the physicist's account of nature as being static and not admitting of the possibility for novelty in the universe. Given the mechanical perspective, all the particles which constitute the universe are ruled by inexorable laws, and their future states could be completely predicted by some Laplacean spirit. Indeed, for Bergson, the mechanistic or physical conception of existence provides no room for the future. There can be no future where all is predetermined. Predicting the future is an illusion because, as Bergson puts it, *tout est donné*.

In the universe which Bergson constructs, however, his starting point is an analysis of the nature of life. There can be no understanding of the world if we look for causes external to the events which constitute reality, regardless of whether these causes are efficient or final. For this reason, Bergson rejects finalism or teleology by stating that teleology is only mechanism turned upside down. To understand the world one must understand the process of life itself—a process which contains within itself a vital force, or *elan vital*, that drives it constantly forward to new forms of existence never before realized. Novelty becomes a reality in the universe of Bergson because it is essentially unpredictable, and it is this inability to determine what the future will yield which so clearly characterize novelty. Living things possess this drive, this impetus, from within. So too, the concept of evolution is adapted not only to the sphere of life, but to that of existence itself. All existence is evolving. The universe is driven by its own dynamism which constantly creates novel modes of being. The nature of evolution as Bergson sees it, and of the vital impetus that drives the evolutionary process, is the subject of the selection included in this text. It provides a brief but vivid glimpse into the heights to which biological insights may be raised. It is the ultimate in biological metaphysics.

More recently there has emerged an approach to living systems which utilizes the insights of both the mechanistic and vitalistic theories but avoids the pitfalls which characterize each one. This perspective is referred to as organismic biology and has as its chief proponent and founder, Ludwig von Bertalanffy. Bertalanffy criticizes the mechanistic conception of life as being inadequate to the task of fully describing the phenomena manifested by living things. The procedure of the mechanist is analysis. The organism is understood by the latter as merely the summation of a number of parts. What is missing in such a characterization of the living system is the organizing principle which adheres between the various parts of subsystems which constitute the total system known as the organism. It is not that mechanism is totally incorrect in its point of view; it is more to the point to say that it fails to capture the fundamental characteristics of the organism because it does not take into account the interrelationships of the various parts. In short, it does not consider the organism as a whole; rather, mechanism sees the living system as an additive combination of parts. It is for this reason that Bertalanffy says, "Analysis of the individual parts and processes in living things is *necessary*, and is the prerequisite for all deeper understandings. Taken alone, however, analysis is not sufficient." The point is, then, that the whole has properties not determined by an addition of the properties of each of the constituent parts. To neglect this fact, argues the organismic biologist, is to miss exactly those features of living systems which render these systems alive rather than not alive. This is not to say that the analysis of various constituent functions and parts of the organism is to be discouraged. On the contrary, biology could not proceed without such an analytical perspective. However, to regard this particular method of investigation as the basis for the interpretation of the nature of living systems is to distort that nature. Organismic biology denies to mechanism the power of ultimate explanation insofar as one is talking about living things.

Vitalism, for all its faults, sees clearly the difficulties of mechanical explanation in the realm of biological phenomena. The solution that is offered, though, provides us with no more aid in getting to the heart of the matter than did the position that it replaces. When vitalism argues that mechanism alone cannot be the mode of biological explanation, it is not rejecting mechanism so much because it is wrong, but because something else is needed. If only the machine were guided by some sort of entelechy, then it would be a living thing. What vitalism does, then, is to add to the machine a guiding force which provides the adaptative and reparative directions for it. With the entelechy, the machine becomes a living machine; but it is still a machine. Driesch adds nothing to mechanism except a non-material entity not subject to the objective observation of scientific procedure. In short, the entelechy becomes a substitute for our ignorance concerning the nature of processes within living systems.

Organismic biology becomes the way out of the dilemma established by the mutual inadequacies of mechanism and vitalism to provide a full and competent theory of life. This is, at least, the attitude of the organismic biologists. Whether or not they succeed in explicating the obscure notion of entelechy so as to provide us with an ultimate basis for explanation, it is not to be doubted that they have contributed a great deal to understanding the nature of living things. By stressing the importance of biological organization and the difference between the properties of the parts of a system and those of the system as a whole, organismic biology seems to avoid the pitfalls of other approaches as well as providing us with a fresh new way to view life and its various processes.

Edmund Sinnott (in the selection in this volume) provides a literary and poignant presentation of the nature and importance of the concept of organization in biology. Sinnott, more than most other thinkers in this book, conveys to the reader the dramatic and profound dimensions of the living system. He is not merely seeking an understanding of life in itself, but instead sees in his search the key to the fundamental nature of existence. The problems of life are ultimate problems and their solutions bring cosmic insights.

Any satisfying philosophy must deal with these questions, and to do so it must be rooted in the science of life itself, of life not only as we see it in man, but as it is expressed in those far simpler organisms up and down the evolutionary scale. It is, therefore, biology in its widest sense, as the interpreter of life at every level, which will bring the richest offerings to philosophy. Tennyson's flower in the crannied wall, if we could really understand it "root and all, and all in all," would indeed solve for us the final mysteries of God and man, for these are the mysteries of life itself.

Since humans as living things formulate questions concerning the nature of the universe, it is natural that at some point they must turn inward in order to understand what the world holds in store for them. There can be no apprehension of the macrocosm without first delving into the depths of microcosmic existence. This is so because the two are intimately related. Just as the Greeks saw the universe imprinted on the soul of man, so Sinnott sees the ultimate questions of human existence impressed in the very matter out of which living things are made.

The selections by Russell, Kapp and Dobbs provide us with several fascinating arguments about some of the issues mentioned above. They are, for all intents and purposes, self-explanatory. The final selection in this section is not clearly related to the rest but is a seminal bit of writing by a great physicist, Erwin Schrödinger. The two chapters from his famous book *What Is Life?* that are included in this volume deal with the problem of the relationship between the science of thermodynamics and biology. Specifically, the major point that is investigated is that of the manner in which the second law of thermodynamics can be reconciled with the observed phenomena of living things. Informally stated, the law holds that in any spontaneous reaction the degree of disorder or *entropy* in the system undergoing the reaction increases. For example, suppose two liquids of different colors are carefully placed in a test tube so that they are separated and share a clear interface with one another. If left alone, these two liquids will ultimately mix, and the distinction between them will no longer be evident because their colors will have combined. Where there was first an ordered arrangement of two differently colored liquids, there is not a single liquid whose molecules are randomly arranged rather than orderly placed. In this system, then, the degree of disorder has increased in the spontaneous reaction whereby the two liquids mixed with one another. In other words, the entropy of the system has increased.

The question then arises as to whether living systems as part of the overall scheme of the universe conform to this second law of thermodynamics. Many observations lead us to think that living systems are exempt from any such obedience. Non-living systems, on the other hand, obey the law very nicely. For example, mountains wear down, rivers widen and their waters run less fiercely, machines wear out. In short, the physical universe seems to be running downhill. But is this the case when we look at the phenomenon of life? The embryo is perhaps the most dramatic example of the way life develops in the opposite direction, of increasing organization and complexity. From a single cell, an enormously involved multicellular organism emerges with a plenitude of interorganizational relations. When living things are challenged, their response seems oriented at maintaining and restoring the maximum degree of order possible. Thus they remain in a dynamic equilibrium with their environment. When that equilibrium is rendered static, the organism is dead.

Do living things, then, actually disobey the second law of thermodynamics? It is left to the reader to examine Schrödinger's selection and determine whether or not a satisfactory answer is given.[1]

While there are many other selections which could be included in this section, it is hoped that the major positions regarding the nature of life have been presented and will stimulate the reader to a fuller investigation of some of the central issues discussed herein.

NOTES

[1]For a more thorough discussion of this problem, see H. F. Blum, *Time's Arrow and Evolution*. London: Oxford University Press, 1955.

Aristotle

Aristotle was born in 384 B.C. in the Greek colony of Stagira in Macedonia. He was the son of Nichomachas, a physician to the royal family of Macedonia. His early training was in medicine, but at the age of seventeen he traveled to Athens and entered the Academy of Plato where he remained until the latter's death in 347 B.C. He spent the next twelve or thirteen years traveling throughout Asia Minor and for three years was tutor to Alexander, then heir to the throne of Macedonia. In 335 B.C. Aristotle returned to Athens and founded the Lyceum, a university of sorts in which Aristotle did the greater part of his biological and philosophical work. When Alexander died in 323 B.C., Aristotle encountered severe opposition to his Macedonian heritage and was forced to flee to Chalcis in Eubcea. He died the next year. Among his most famous works are: *Nicomachean Ethics, Metaphysics, Politics, Poetics, On the Parts of Animals, On the Reproduction of Animals,* and *On the Soul.*

The following selection is taken from Book II of *Aristotle's Psychology*, Edwin Wallace's translation of *De Anima* published by Cambridge University Press in 1882.

DE ANIMA

Book Second
Chapter I

The psychological theories of earlier thinkers have occupied us hitherto. We will now take up the subject as it were afresh, and attempt to determine what soul is, and what is the most comprehensive definition that can be given of it.

Real substance is the name which we assign one class of existing things; and this real substance may be viewed from several aspects, either, *firstly*, as matter, meaning by matter that which in itself is not any individual thing; or *secondly*, as form and specific characteristic in virtue of which an object comes to be described as such and such an individual; or *thirdly*, as the result produced by a combination of this matter and this form. Further, while matter is merely potential existence, the form is perfect realization (a conception which may be taken in two forms, either as resembling knowledge possessed or as corresponding to observation in active exercise).

These real substances again are thought to correspond for the most part with bodies, and more particularly with natural bodies, because these latter are the source from which other bodies are formed. Now among such natural bodies, some have, others do not have life, meaning here by life the process of nutrition, increase and decay from an internal principle. Thus every natural body possessed of life would be a real substance, and a substance which we may describe as composite.

Since then the body, as possessed of life, is of this compound character, the body itself would not constitute the soul: for body is not [like life and soul] something attributed to a subject; it rather acts as the underlying subject and the material basis. Thus then the soul must necessarily be a real substance, as the form which determines a natural body possessed potentially of life. The reality however of an object is contained in its perfect realization. Soul therefore will be a perfect realization of a body such as has been described. Perfect realization however is a word used in two senses: it may be understood either as an implicit state corresponding to knowledge as possessed, or as an explicitly exercised process corresponding to active observation. Here, in reference to soul, it must evidently be understood in the former of these two senses: for the soul is present with us as much while we are asleep as while we are awake; and while waking resembles active observation, sleep resembles the implicit though not exercised possession of knowledge. Now in reference to the same subject, it is the implicit knowledge of scientific principles which stands prior. Soul therefore is the earlier or implicit perfect realization of a natural body possessed potentially of life.

Such potential life belongs to everything which is possessed of organs. Organs however, we must remember, is a name that applies also to the parts of plants, except that they are altogether uncompounded. Thus the leaf is the protection of the pericarp and the pericarp of the fruit; while the roots are analogous to the mouth in animals, both being used to absorb nourishment. Thus then, if we be required to frame some one common definition, which will apply to every form of soul, it would be that soul is the earlier perfect realization of a natural organic body.

The definition we have just given should make it evident that we must no more ask whether the soul and the body are one, than ask whether the wax and the figure impressed upon it are one, or generally inquire whether the material and that of which it is the material are one; for though unity and being are used in a variety of senses, their most distinctive sense is that of perfect realization.

A general account has thus been given of the nature of the soul: it is, we have seen, a real substance which expresses an idea. Such a substance is the manifestation of the inner meaning of such and such a body. Suppose, for example, that an instrument such as an axe were a natural body: then its axehood or its being an axe would constitute its essential nature or reality, and thus, so to speak, its soul; because were this axehood taken away from it, it would be no longer an axe, except in so far as it might still be called by this same name. The object in question, however, is as matter of fact only an axe; soul being not the idea and the manifestation of the meaning of a body of this kind, but of a natural body possessing within itself a cause of movement and of rest.

The theory just stated should be viewed also in reference to the separate bodily parts. If, for example, the eye were possessed of life, vision would be its soul: because vision is the reality which expresses the idea of the eye. The eye itself, on the other hand, is merely the material substratum for vision: and when this power of vision fails, it no longer remains an eye, except in so far as it is still called by the same name, just in the same way as an eye carved in stone or delineated in painting is also so described. Now what holds good of the part must be applied to the living body taken as a whole: for perception as a whole stands to the whole sensitive body, as such, in the same ratio as the particular exercise of sense stands to a single organ of sense.

The part of our definition which speaks of something as "potentially possessed of life" must be taken to mean not that which has thrown off its soul, but rather that which has it: the seed and the fruit is such and such a body potentially. In the same way then as cutting is the full realization of an axe, or actual seeing the realization of the eye, so also waking may be said to be the full realization of the body: but it is in the sense in which vision is not only the exercise but also the implicit capacity of the eye that soul is the true realization of the body. The body on the other hand is merely the material to which soul gives reality: and just as the eye is both the pupil and its vision, so also the living animal is at once the soul and body in connection.

It is not then difficult to see that soul or certain parts of it (if it naturally admit of partition) cannot be separated from the body: for in some cases the soul is the realization of the parts of body themselves. It is however perfectly conceivable that there may be some parts of it which are separable and this because they are not the expression or realization of any particular body. And indeed it is further matter of doubt whether soul as the perfect realization of the body may not stand to it in the same separable relation as a sailor to his boat.

This much may suffice as a description and sketch of the nature of the soul.

Chapter II

It is however by proceeding from that which in the order of nature is indistinct, but is relatively to us more obvious and manifest, that we reach what is clear and more intelligible in the order of thought. We must therefore make a fresh attempt to discuss soul in this manner. For a definition should not, as most definitions do, merely assert the existence of an object and say what it is: it should also contain and express the cause or reason of the object. But, as usually framed, the terms of definitions are merely like conclusions. Thus, for example, let us ask— What is squaring? Squaring, it will be answered, is the construction of a rectangular equilateral figure equal to another figure with unequal sides. Now such a definition is merely like the statement of a conclusion. To say, on the other hand, that squaring is the discovery of a mean proportional is to state the cause which explains the result.

It may serve as a fresh beginning for our inquiry to say that the animate is distinguished from the inanimate or soulless by the fact of life. There are a number of ways in which a thing is said to live; yet should it possess only one of them—as for example, reason, sense—perception, local movement and rest, and further movement in respect of nutrition as well as of decay and growth—we say it lives. Hence it is that all plants are thought to live; because they manifestly contain within themselves such a power and principle as enables them to acquire growth and undergo decay in opposite directions; for they do not while growing upwards not grow downwards but they grow in both directions and on all sides, and they continue to live so long as they can assimilate nourishment. Now this faculty of nutrition may be separated from the other functions; but in the case of mortal creatures the other faculties cannot exist apart from this, as indeed is evident from plants which possess no other psychic power except this faculty of growth.

It is then through this principle of nutrition that life is an attribute of all living things. At the same time the animal strictly so called only begins when we reach sensation: for even those objects which do not move themselves nor change their position but possess sensation are said to be animals and not merely to be living. Among the senses themselves, it is touch which is the fundamental attribute of all animal forms. And just as the nutritive function may exist apart from touch and every form of sense, so also may touch exist without any of the other senses. Thus while nutritive is the name given to that part of the soul in which plants share as well as animals, all animals are found to possess the sense of touch. Why each of these faculties is so allotted we shall state hereafter: here it may be enough to say that the soul is the source and centre of the various states here mentioned and is determined and defined by those powers of nutrition, sensation, understanding and movement.

With regard to these several functions, whether each is the soul or a part of the soul; and if a part, whether so as only to be separable in thought or actually in space—with regard to some of these questions it is not difficult to see the answer, while others present difficulties. For just as, in the case of plants, some parts when divided are found to live even when separated from one another—a fact which seems to show that the soul within them exists as actually one though it is potentially several; so also do we see it happen with respect to another specific aspect of the soul in the case of insects which have been divided. In such a case, each of the divided parts possesses sensation and the power of local movement, and if sensation, then also in addition imagination and desire: for where sense is present, there pain and pleasure follow also as concomitants, and where pain and pleasure exist, appetite is also necessarily present. With regard on the other hand to reason and the faculty of thought we have as yet no obvious facts to appeal to. Reason however would seem to constitute a different phase of soul from those we have already noticed and it alone admits of separation as the eternal from the perishable. But as for the other parts of soul, it is clear from these considerations that they are not separated in the way that some maintain. At the same time it is evident that in thought and by abstraction they may be divided from one another. The sensitivity is one thing, the reflective faculty another, if it be one thing to have sensation, another thing to exercise reflection. And this same truth holds good also of the other powers which have been described.

Respecting these various powers, there are some animals which possess them all, others which have merely some of them, and others again which have but one only. It is this which makes the difference between one class of animals and another, though the reason for this fact can only be investigated afterwards. The same thing may be noticed also as regards the senses. Some animals have all of them, others have but some, and a third class possesses only that one sense which is most indispensable—viz. touch.

[Life, then, and sensation are what mark the animate.] But there are two ways in which we may speak of that by which we live and have sensation just as also that by which we know may be employed to denote either knowledge or the mind, by both of which we are in the habit of speaking of people as knowing. So also that by which we are in health denotes on the one hand the health itself, on the other hand some portion of the body or it may be the whole of it. Now of these two uses, knowledge and health are what we may term the determining form and notion and so to speak the realization of the recipient faculty, in the one case of knowledge, in the other of health—for the passive material which is subject to modification is what is taken to be the home of the manifestation of the active forces. Soul then is the original and fundamental ground of all our life, of our sensation and of our reasoning. It follows therefore that the soul must be regarded as a sort of form and idea, rather than as matter and as underlying subject. For the term real substance is, as we have before remarked, employed in three senses: it may denote either the specific form, or the material substratum, or thirdly the combination of the two: and of these different aspects of reality the matter or substratum is but the potential ground, whereas the form is the perfect realization. Since then it is the product of the two that is animate, it cannot be that the body is the full realization or expression of the soul; rather on the contrary it is the soul which is the full realization of some body.

This fact fully supports the view of those who hold that the soul is not independent of some sort of body and yet not to be identified with a body of any sort whatever. The truth is that soul is not body but it is something which belongs to body. And hence further it exists in a body and in a body of such and such a nature, not left undetermined in the way that earlier thinkers introduced it into the body without determining besides what and what sort of body it was, although it does not even look as though any casual thing admitted any other casual thing.

This same conclusion may be reached also on *a priori* grounds. The full realization of each object is naturally reached only within that which is potentially existent and within that material substratum which is appropriate to it. It is clear then from these considerations that soul is a kind of full realization or expression of the idea of that which has potentially the power to be of such a character.

René Descartes

René Descartes, who was born in 1596 of wealthy parents, is often viewed as the father of modern philosophy. He lived in Brittany where he was nurtured by the Jesuits. Several years were spent in Paris where he worked as an engineer officer. Eventually, Descartes moved to Holland where he was free from the influence of the Catholic Church and completed his most important work there. He is best known for his *Meditations, Discourse on Method, The Passions of the Soul*, and *The Principles of Philosophy*. He died in Stockholm in 1650, a few months after having traveled to Sweden at Queen Christina's personal request to be her instructor in philosophy.

The following selection is taken from *The Philosophical Works of Descartes*, Vol. I, translated by Elizabeth S. Haldane and G. R. T. Ross and published by Cambridge University Press, 1911-1912.

THE PASSIONS OF THE SOUL

Part First

Of the Passions in General, and Incidentally of the Whole Nature of Man

Article I

That what in respect of a subject is passion, is in some other regard always action.

There is nothing in which the defective nature of the sciences which we have received from the ancients appears more clearly than in what they have written on the passions; for, although this is a matter which has at all times been the object of much investigation, and though it would not appear to be one of the most difficult, inasmuch as since every one has experience of the passions within himself, there is no necessity to borrow one's observations from elsewhere in order to discover their nature: yet that which the ancients have taught regarding them is both so slight, and for the most part so far from credible, that I am unable to entertain any hope of approximating to the truth excepting by shunning the paths which they have followed. This is why I shall be here obliged to write just as though I were treating of a matter which no one had ever touched on before me; and, to begin with, I consider that all that which occurs or that happens anew, is by the philosophers, generally speaking, termed a passion, in as far as the subject to which it occurs is concerned, and an action in respect of him who causes it to occur. Thus although the agent and the recipient [patient] are frequently very different, the action and the passion are always one and the same thing, although having different names, because of the two diverse subjects to which it may be related.

Article II

That in order to understand the passions of the soul its functions must be distinguished from those of body.

Next I note also that we do not observe the existence of any subject which more immediately acts upon our soul than the body to which it is joined, and that we must consequently consider that what in the soul is a passion is in the body commonly speaking an action; so that there is no better means of arriving at a knowledge of our passions than to examine the difference which exists between soul and body in order to know to which of the two we must attribute each one of the functions which are within us.

Article III

What rule we must follow to bring about this result.

As to this we shall not find much difficulty if we realise that all that we experience as being in us, and that to observation may exist in wholly inanimate bodies, must be attributed to our body alone; and, on the other hand, that all that which is in us and which we cannot in any way conceive as possibly pertaining to a body, must be attributed to our soul.

Article IV

That the heat and movement of the members proceed from the body, the thoughts from the soul.

Thus because we have no conception of the body as thinking in any way, we have reason to believe that every kind of thought which exists in us belongs to the soul; and because we do not doubt there being inanimate bodies which can move in as many as or in more diverse modes than can ours, and which have as much heat or more (experience demonstrates this to us in flame, which of itself has much more heat and movement than any of our members), we must believe that all the heat and all the movements which are in us pertain only to body, inasmuch as they do not depend on thought at all.

Article V

That it is an error to believe that the soul supplies the movement and heat to body.

By this means we shall avoid a very considerable error into which many have fallen; so much so that I am of opinion that this is the primary cause which has prevented our being able hitherto satisfactorily to explain the passions and the other properties of the soul. It arises from the fact that from observing that all dead bodies are devoid of heat and consequently of movement, it has been thought that it was the absence of soul which caused these movements and this heat to cease; and thus, without any reason, it was thought that our natural heat and all the movements of our body depend on the soul: while in fact we ought on the contrary to believe that the soul quits us on death only because this heat ceases, and the organs which serve to move the body disintegrate.

Article VI

The difference that exists between a living body and a dead body.

In order, then, that we may avoid this error, let us consider that death never comes to pass by reason of the soul, but only because some one of the principal parts of the body decays; and we may judge that the body of a living man differs from that of a dead man just as does a watch or other automaton (i.e. a machine that moves of itself), when it is wound up and contains in itself the corporeal principle of those movements for which it is designed along with all that is requisite for its action, from the same watch or other machine when it is broken and when the principle of its movement ceases to act.

Article VII

A brief explanation of the parts of the body and some of its functions.

In order to render this more intelligible, I shall here explain in a few words the whole method in which the bodily machine is composed. There is no one who does not already know that there are in us a heart, a brain, a stomach, muscles, nerves, arteries, veins, and such things. We also know that the food that we eat descends into the stomach and bowels where its juice, passing into the liver and into all the veins, mingles with, and thereby increases the quantity of the blood which they contain. Those who have acquired even the minimum of medical knowledge further know how the heart is composed, and how all the blood in the veins can easily flow from the vena cava into its right side and from thence pass into the lung by the vessel which we term the arterial vein, and then return from the lung into the left side of the heart, by the vessel called the venous artery, and finally pass from there into the great artery, whose branches spread throughout all the body. Likewise all those whom the authority of the ancients has not entirely blinded, and who have chosen to open their eyes for the purpose of investigating the opinion of Harvey regarding the circulation of the blood, do not doubt that all the veins and arteries of the body are like streams by which the blood ceaselessly flows with great swiftness, taking its course from the right cavity of the heart by the arterial vein whose branches are spread over the whole of the lung, and joined to that of the venous artery by which it passes from the lung into the left side of the heart; from these, again, it goes into the great artery whose branches, spread throughout all the rest of the body, are united to the branches of the vein, which branches once more carry the same blood into the right cavity of the heart. Thus these two cavities are like sluices through each of which all the blood passes in the course of each circuit which it makes in the body. We further know that all the movements of the members depend on the muscles, and that these muscles are so mutually related one to another that when the one is contracted it draws toward itself the part of the body to which it is attached, which causes the opposite muscle at the same time to become elongated; then if at another time it happens that this last contracts, it causes the former to become elongated and it draws back to itself the part to which they are attached. We know finally that all these movements of the muscles, as also all the senses, depend on the nerves, which resemble small fila- ments, or little tubes, which all proceed from the brain, and thus contain like it a certain very subtle air or wind which is called the animal spirits.

Article VIII

What is the principle of all these functions?
But it is not usually known in what way these animal spirits and these nerves contribute to the movements and to the senses, nor what is the corporeal principle which causes them to act. That is why, although I have already made some mention of them in my other writings, I shall not here omit to say shortly that so long as we live there is a continual heat in our heart, which is a species of fire which the blood of the veins there maintains, and that this fire is the corporeal principle of all the movements of our members.

Article IX

How the movement of the heart is carried on.
Its first effect is to dilate the blood with which the cavities of the heart are filled; that causes this blood, which requires a greater space for its occupation, to pass impetuously from the right cavity into the arterial vein, and from the left into the great artery; then when this dilation ceases, new blood immediately enters from the vena cava into the right cavity of the heart, and from the venous artery into the left; for there are little membranes at the entrances of these four vessels, disposed in such a manner that they do not allow the blood to enter the heart but by the two last, nor to issue from it but by the two others. The new blood which has entered into the heart is then immediately afterwards rarefied, in the same manner as that which preceded it; and it is just this which causes the pulse, or beating of the heart and arteries; so that this beating repeats itself as often as the new blood enters the heart. It is also just this which gives its motion to the blood, and causes it to flow ceaselessly and very quickly in all the arteries and veins, whereby it carries the heat

which it acquires in the heart to every part of the body, and supplies them with nourishment.

Article X

How the animal spirits are produced in the brain.

But what is here most worthy of remark is that all the most animated and subtle portions of the blood which the heat has rarefied in the heart, enter ceaselessly in large quantities into the cavities of the brain. And the reason which causes them to go there rather than elsewhere, is that all the blood which issues from the heart by the great artery takes its course in a straight line towards that place, and not being able to enter it in its entirety, because there are only very narrow passages there, those of its parts which are the most agitated and the most subtle alone pass through, while the rest spreads abroad in all the other portions of the body. But these very subtle parts of the blood form the animal spirits; and for this end they have no need to experience any other change in the brain, unless it be that they are separated from the other less subtle portions of the blood; for what I here name spirits are nothing but material bodies and their one peculiarity is that they are bodies of extreme minuteness and that they move very quickly like the particles of the flame which issues from a torch. Thus it is that they never remain at rest in any spot, and just as some of them enter into the cavities of the brain, others issue forth by the pores which are in its substance, which pores conduct them into the nerves, and from there into the muscles, by means of which they move the body in all the different ways in which it can be moved. . . .

Article XXVII

The definition of the passions of the soul.

After having considered in what the passions of the soul differ from all its other thoughts, it seems to me that we may define them generally as the perceptions, feelings, or emotions of the soul which we relate specially to it, and which are caused, maintained, and fortified by some movement of the spirits.

Article XXVIII

Explanation of the first part of this definition.

We may call them perceptions when we make use of this word generally to signify all the thoughts which are not actions of the soul, or desires, but not when the term is used only to signify clear cognition; for experience shows us that those who are the most agitated by their passions, are not those who know them best; and that they are of the number of perceptions which the close alliance which exists between the soul and the body, renders confused and obscure. We may also call them feelings because they are received into the soul in the same way as are the objects of our outside senses, and are not otherwise known by it; but we can yet more accurately call them emotions of the soul, not only because the name may be attributed to all the changes which occur in it—that is, in all the diverse thoughts which come to it, but more especially because of all the kinds of thought which it may have, there are no others which so powerfully agitate and disturb it as do these passions.

Article XXIX

Explanation of the second part.

I add that they particularly relate to the soul, in order to distinguish them from the other feelings which are related, the one to outside objects such as scents, sounds, and colours; the others to our body such as hunger, thirst, and pain. I also add that they are caused, maintained, and fortified by some movement of the spirits, in order to distinguish them from our desires, which we may call emotions of the soul which relate to it, but which are caused by itself; and also in order to explain their ultimate and most proximate cause, which plainly distinguishes them from the other feelings.

Article XXX

That the soul is united to all the portions of the body conjointly.

But in order to understand all these things more perfectly, we must know that the soul is really joined to the whole body, and that we cannot, properly speaking, say that it exists in any one of its parts to the exclusion of the others, because it is one and in some manner indivisible, owing to the disposition of its organs, which are so related to one another that when any one of them is removed, that renders the whole body defective; and because it is of a nature which has no relation to extension, nor dimensions, nor other properties of the matter of which the body is composed, but only to the whole conglomerate of its organs, as appears from the fact that we could not in any way conceive of the half or the third of a soul, nor of the space it occupies, and because it does not become smaller owing to the cutting off of some portion of the body, but separates itself from it entirely when the union of its assembled organs is dissolved.

Article XXXI

That there is a small gland in the brain in which the soul exercises its functions more particularly than in the other parts.

It is likewise necessary to know that although the soul is joined to the whole body, there is yet in that a certain part in which it exercises its functions more particularly than in all the others; and it is usually believed that this part is the brain, or possibly the heart: the brain, because it is with it that the organs of sense are connected, and the heart because it is apparently in it that we experience the passions. But, in examining the matter with care, it seems as though I had clearly ascertained that the part of the body in which the soul exercises its functions immediately is in nowise the heart, nor the whole of the brain, but merely the most inward of all its parts, to wit, a certain very small gland which is situated in the middle of its substance and so suspended above the duct whereby the animal spirits in its anterior cavities have communication with those in the posterior, that the slightest movements which take place in it may alter very greatly the course of these spirits; and reciprocally that the smallest changes which occur in the course of the spirits may do much to change the movements of this gland.

Article XXXII

How we know that this gland is the main seat of the soul.

The reason which persuades me that the soul cannot have any other seat in all the body than this gland wherein to exercise its functions immediately, is that I reflect that the other parts of our brain are all of them double, just as we have two eyes, two hands, two ears, and finally all the organs of our outside senses are double; and inasmuch as we have but one solitary and simple thought of one particular thing at one and the same moment, it must necessarily be the case that there must somewhere be a place where the two images which come to us by the two eyes, where the two other impressions which proceed from a single object by means of the double organs of the other senses, can unite before arriving at the soul, in order that they may not represent to it two objects instead of one. And it is easy to apprehend how these images or other impressions might unite in this gland by the intermission of the spirits which fill the cavities of the brain; but there is no other place in the body where they can be thus united unless they are so in this gland.

Article XXXIII

That the seat of the passions is not in the heart.

As to the opinion of those who think that the soul receives its passions in the heart, it is not of much consideration, for it is only founded on the fact that the passions cause us to feel some change taking place there; and it is easy to see that this change is not felt in the heart excepting through the medium of a small nerve which descends from the brain towards it, just as pain is felt as in the foot by means of the nerves of the foot, and the stars are perceived as in the heavens by means of their light and of the optic nerves; so that it is not more necessary that our soul should exercise its functions immediately in the heart, in order to feel its passions there, than it is necessary for the soul to be in the heavens in order to see the stars there.

Article XXXIV

How the soul and the body act on one another.

Let us then conceive here that the soul has its principal seat in the little gland which exists in the middle of the brain, from whence it radiates forth through all the remainder of the body by means of the animal spirits, nerves, and even the blood, which, participating in the impressions of the spirits, can carry them by the arteries into all the

members. And recollecting what has been said above about the machine
of our body, i.e. that the little filaments of our nerves are so distributed
in all its parts, that on the occasion of the diverse movements which are
there excited by sensible objects, they open in diverse ways the pores of
the brain, which causes the animal spirits contained in these cavities to
enter in diverse ways into the muscles, by which means they can move
the members in all the different ways in which they are capable of being
moved; and also that all the other causes which are capable of moving
the spirits in diverse ways suffice to conduct them into diverse muscles;
let us here add that the small gland which is the main seat of the soul is so
suspended between the cavities which contain the spirits that it can be
moved by them in as many different ways as there are sensible diver-
sities in the object, but that it may also be moved in diverse ways by the
soul, whose nature is such that it receives in itself as many diverse
impressions, that is to say, that it possesses as many diverse perceptions
as there are diverse movements in this gland. Reciprocally, likewise, the
machine of the body is so formed that from the simple fact that this gland
is diversely moved by the soul, or by such other cause, whatever it is, it
thrusts the spirits which surround it towards the pores of the brain,
which conduct them by the nerves into the muscles, by which means it
causes them to move the limbs.

Article XXXV

*Example of the mode in which the impressions of the objects unite in
the gland which is in the middle of the brain.*

Thus, for example, if we see some animal approach us, the light
reflected from its body depicts two images of it, one in each of our eyes,
and these two images form two others, by means of the optic nerves, in
the interior surface of the brain which faces its cavities; then from there,
by means of the animal spirits with which its cavities are filled, these
images so radiate towards the little gland which is surrounded by these
spirits, that the movement which forms each point of one of the images
tends towards the same point of the gland towards which tends the
movement which forms the point of the other image, which represents
the same part of this animal. By this means the two images which are in
the brain form but one upon the gland, which, acting immediately upon
the soul, causes it to see the form of this animal.

Article XXXVI

Example of the way in which the passions are excited in the soul.

And, besides that, if this figure is very strange and frightful—that is, if it has a close relationship with the things which have been formerly hurtful to the body, that excites the passion of apprehension in the soul and then that of courage, or else that of fear and consternation according to the particular temperament of the body or the strength of the soul, and according as we have to begin with been secured by defence or by flight against the hurtful things to which the present impression is related. For in certain persons that disposes the brain in such a way that the spirits reflected from the image thus formed on the gland, proceed thence to take their places partly in the nerves which serve to turn the back and dispose the legs for flight, and partly in those which so increase or diminish the orifices of the heart, or at least which so agitate the other parts from whence the blood is sent to it, that this blood being there rarefied in a different manner from usual, sends to the brain the spirits which are adapted for the maintenance and strengthening of the passion of fear, i.e. which are adapted to the holding open, or at least reopening, of the pores of the brain which conduct them into the same nerves. For from the fact alone that these spirits enter into these pores, they excite a particular movement in this gland which is instituted by nature in order to cause the soul to be sensible of this passion; and because these pores are principally in relation with the little nerves which serve to contract or enlarge the orifices of the heart, that causes the soul to be sensible of it for the most part as in the heart.

Article XXXVII

How it seems as though they are all caused by some movement of the spirits.

And because the same occurs in all the other passions, to wit, that they are principally caused by the spirits which are contained in the cavities of the brain, inasmuch as they take their course towards the nerves which serve to enlarge or contract the orifices of the heart, or to drive in various ways to it the blood which is in the other parts, or, in whatever other fashion it may be, to carry on the same passion, we may from this clearly understand why I have placed in my definition of them above, that they are caused by some particular movement of the animal spiriṭs.

Article XXXVIII

Example of the movements of the body which accompany the passions and do not depend on the soul.

For the rest, in the same way as the course which these spirits take towards the nerves of the heart suffices to give the movement to the gland by which fear is placed in the soul, so, too, by the simple fact that certain spirits at the same time proceed towards the nerves which serve to move the legs in order to take flight, they cause another movement in the same gland, by means of which the soul is sensible of and perceives this flight, which in this way may be excited in the body of the disposition of the organs alone, and without the soul's contributing thereto.

Article XXXIX

How one and the same cause may excite different passions in different men.

The same impression which a terrifying object makes on the gland, and which causes fear in certain men, may excite in others courage and confidence; the reason of this is that all brains are not constituted in the same way, and that the same movements of the gland which in some excites fear, in others causes the spirits to enter into the pores of the brain which conduct them partly into the nerves which serve to move the hands for purposes of self-defence, and partly into those which agitate and drive the blood towards the heart in the manner requisite to produce the spirits proper for the continuance of this defence, and to retain the desire of it.

Article XL

The principal effect of the passions.

For it is requisite to notice that the principal effect of all the passions in men is that they incite and dispose their soul to desire those things for which they prepare their body, so that the feeling of fear incites it to desire to fly, that of courage to desire to fight, and so on.

Article XLI

The power of the soul in regard to the body.

But the will is so free in its nature, that it can never be constrained; and of the two sorts of thoughts which I have distinguished in the soul (of which the first are its actions, i.e. its desires, the others its passions, taking this word in its most general significance, which comprises all kinds of perceptions), the former are absolutely in its power, and can only be indirectly changed by the body, while on the other hand the latter depend absolutely on the actions which govern and direct them, and they can only indirectly be altered by the soul, excepting when it is itself their cause. And the whole action of the soul consists in this, that solely because it desires something, it causes the little gland to which it is closely united to move in the way requisite to produce the effect which relates to this desire.

Julien Offrey de La Mettrie

Born in Brittany in 1709 as the son of a wealthy merchant, Julien Offrey de La Mettrie was originally destined to become a priest. He studied theology in Paris but was ultimately persuaded to study medicine. This he did at first in Paris and then at Leyden. After passing his examinations, La Mettrie became physician to a military regiment in Paris and soon encountered the enmity of his fellow physicians because of his unorthodox views. When he published his now famous treatise *The Natural History of the Soul*, he was regarded as thoroughly opposed to many of the views of the Catholic Church. He was advised to flee to Holland, but while living in Leyden he persisted in his unorthodox pursuits and published a poorly disguised anonymous pamphlet, *Man a Machine*. Even the Dutch could not be relied upon to sanction this almost heretical work, and La Mettrie was most fortunate in being invited to the Court of Frederick II of Prussia, a monarch whose concerns about religious disputes were minimal. La Mettrie held the post of lecturer at the royal court and also practiced as a physician. He died apparently of food poisoning in 1751.

The following selection is taken from *Man a Machine*, pp. 140-49, translated by Gertrude C. Bussey and revised by M. W. Calkins and published by the Open Court Publishing Company in 1912.

MAN A MACHINE

Grant only that organized matter is endowed with a principle of motion, which alone differentiates it from the inorganic (and can one deny this in the face of the most incontestable observation?) and that among animals, as I have sufficiently proved, everything depends upon the diversity of this organization: these admissions suffice for guessing the riddle of substances and of man. It [thus] appears that there is but one [type of organization] in the universe, and that man is the most perfect [example]. He is to the ape, and to the most intelligent animals, as the planetary pendulum of Huyghens is to a watch of Julien Leroy. More instruments, more wheels and more springs were necessary to mark the movements of the planets than to mark or strike the hours; and Vaucanson, who needed more skill for making his flute player than for making his duck, would have needed still more to make a talking man, a mechanism no longer to be regarded as impossible, especially in the hands of another Prometheus. In like fashion, it was necessary that nature should use more elaborate art in making and sustaining a machine which for a whole century could mark all motions of the heart and of the mind; for though one does not tell time by the pulse, it is at least the barometer of the warmth and the vivacity by which one may estimate the nature of the soul. I am right! The human body is a watch, a large watch constructed with such skill and ingenuity, that if the wheel which marks the seconds happens to stop, the minute wheel turns and keeps on going its round, and in the same way the quarter-hour wheel, and all the others go on running when the first wheels have stopped because rusty or, for any reason, out of order. Is it not for a similar reason that the stoppage of a few blood vessels is not enough to destroy or suspend the strength of the movement which is in the heart as in the mainspring of the machine; since, on the contrary, the fluids whose volume is diminished, having a shorter road to travel, cover the ground more quickly, borne on as by a fresh current which the energy of the heart increases in proportion to the resistance it encounters at the ends of the blood-vessels? And is not this the reason why the loss of sight (caused by the compression of the optic nerve and by its ceasing to convey the images of objects) no more hinders hearing, than the loss of hearing (caused by obstruction of the functions of the auditory nerve) implies the loss of sight? In the same way, finally, does not one man hear (except immediately after his attack) without being able to say that he hears, while another who hears nothing, but whose lingual nerves are uninjured in the brain, mechanically tells of all the dreams which pass through his mind? These phenomena do not surprise enlightened physicians at all. They know what to think about man's nature, and (more accurately to express myself in passing) of two physicians, the better one and the one who deserves more confidence is always, in my opinion, the one who is more versed in the physique or mechanism of the human body, and who, leaving aside the soul and all the anxieties which this chimera gives to fools and to ignorant men, is seriously occupied only in pure naturalism.

Therefore let the pretended M. Charp deride philosophers who have regarded animals as machines. How different is my view! I believe that Descartes would be a man in every way worthy of respect, if, born in a century that he had not been obliged to enlighten, he had known the value of experiment and observation, and the danger of cutting loose from them. But it is none the less just for me to make an authentic reparation to this great man for all the insignificant philosophers—poor jesters, and poor imitators of Locke—who instead of laughing impudently at Descartes, might better realize that without him the field of philosophy, like the field of science without Newton, might perhaps be still uncultivated.

This celebrated philosopher, it is true, was much deceived, and no one denies that. But at any rate he understood animal nature, he was the first to prove completely that animals are pure machines. And after a discovery of this importance demanding so much sagacity, how can we without ingratitude fail to pardon all his errors!

In my eyes, they are all atoned for by that great confession. For after all, although he extols the distinctness of the two substances, this is plainly but a trick of skill, a ruse of style, to make theologians swallow a poison, hidden in the shade of an analogy which strikes everybody else and which they alone fail to notice. For it is this, this strong analogy, which forces all scholars and wise judges to confess that these proud and vain beings, more distinguished by their pride than by the name of men however much they may wish to exalt themselves, are at bottom only animals and machines which, though upright, go on all fours. They all have this marvelous instinct, which is developed by education into mind, and which always has its seat in the brain, (or for want of that when it is lacking or hardened, in the medulla oblongata) and never in the cerebellum; for I have often seen the cerebellum injured, and other observers[1] have found it hardened, when the soul has not ceased to fulfil its functions.

To be a machine, to feel, to think, to know how to distinguish good from bad, as well as blue from yellow, in a word, to be born with an intelligence and a sure moral instinct, and to be but an animal, are therefore characters which are no more contradictory, than to be an ape or a parrot and to be able to give oneself pleasure. . . . I believe that thought is so little incompatible with organized matter, that it seems to be one of its properties on a par with electricity, the faculty of motion, impenetrability, extension, etc.

Do you ask for further observations? Here are some which are incontestable and which all prove that man resembles animals perfectly, in his origin as well as in all the points in which we have thought it essential to make the comparison. . . .

Let us observe man both in and out of his shell, let us examine young embryos of four, six, eight or fifteen days with a microscope; after that time our eyes are sufficient. What do we see? The head alone; a little round egg with two black points which mark the eyes. Before that, everything is formless, and one sees only a medullary pulp, which is the brain, in which are formed first the roots of the nerves, that is, the principle of feeling, and the heart, which already within this substance has the power of beating of itself; it is the *punctum saliens* of Malpighi, which perhaps already owes a part of its excitability to the influence of the nerves. Then little by little, one sees the head lengthen from the neck, which, in dilating, forms first the thorax inside which the heart has already sunk, there to become stationary; below that is the abdomen which is divided by a partition (the diaphragm). One of these enlargements of the body forms the arms, the hands, the fingers, the nails, and the hair; the other forms the thighs, the legs, the feet, etc., which differ only in their observed situation, and which constitute the support and the balancing pole of the body. The whole process is a strange sort of growth, like that of plants. On the tops of our heads is hair in place of which the plants have leaves and flowers; everywhere is shown the same luxury of nature, and finally the directing principle of plants is placed where we have our soul, that other quintessence of man.

Such is the uniformity of nature, which we are beginning to realize; and the analogy of the animal with the vegetable kingdom, of man with plant. Perhaps there even are animal plants, which in vegetating, either fight as polyps do, or perform other functions characteristic of animals.

We are veritable moles in the field of nature; we achieve little more than the mole's journey and it is our pride which prescribes limits to the limitless. We are in the position of a watch that should say (a writer of fables would make the watch a hero in a silly tale): "I was never made by that fool of a workman, I who divide time, who mark so exactly the course of the sun, who repeat aloud the hours which I mark! No! that is impossible!" In the same way, we disdain, ungrateful wretches that we are, this common mother of all kingdoms, as the chemists say. We imagine, or rather we infer, a cause superior to that to which we owe all, and which truly has wrought all things in an inconceivable fashion. No;

matter contains nothing base, except to the vulgar eyes which do not recognize her in her most splendid works; and nature is no stupid workman. She creates millions of men, with a facility and a pleasure more intense than the effort of a watchmaker in making the most complicated watch. Her power shines forth equally in creating the lowliest insect and in creating the most highly developed man; the animal kingdom costs her no more than the vegetable, and the most splendid genius no more than a blade of wheat. Let us then judge by what we see of that which is hidden from the curiosity of our eyes and of our investigations, and let us not imagine anything beyond. Let us observe the ape, the beaver, the elephant, etc., in their operations. If it is clear that these activities can not be performed without intelligence, why refuse intelligence to these animals? And if you grant them a soul, you are lost, you fanatics! You will in vain say that you assert nothing about the nature of the animal soul and that you deny its immortality. Who does not see that this is a gratuitous assertion; who does not see that the soul of an animal must be either mortal or immortal, whichever ours [is], and that it must therefore undergo the same fate as ours, whatever that may be, and that thus [in admitting that animals have souls], you fall into Scylla in the effort to avoid Charybdis?

Break the chain of your prejudices, arm yourselves with the torch of experience, and you will render to nature the honor she deserves, instead of inferring anything to her disadvantage, from the ignorance in which she has left you. Only open wide your eyes, only disregard what you can not understand, and you will see that the ploughman whose intelligence and ideas extend no further than the bounds of his furrow, does not differ essentially from the greatest genius,—a truth which the dissection of Descartes's and of Newton's brains would have proved; you will be persuaded that the imbecile and the fool are animals with human faces, as the intelligent ape is a little man in another shape; in short, you will learn that since everything depends absolutely on difference of organization, a well constructed animal which has studied astronomy, can predict an eclipse, as it can predict recovery or death when it has used its genius and its clearness of vision, for a time, in the school of Hippocrates and at the bedside of the sick. By this line of observations and truths, we come to connect the admirable power of thought with matter, without being able to see the links, because the subject of this attribute is essentially unknown to us.

Let us not say that every machine or every animal perishes altogether or assumes another form after death, for we know absolutely nothing about the subject. On the other hand, to assert that an immortal machine is a chimera or a logical fiction, is to reason as absurdly as caterpillars would reason if, seeing the cast-off skins of their fellow-caterpillars, they should bitterly deplore the fate of their species, which to them would seem to come to nothing. The soul of these insects (for each animal has his own) is too limited to comprehend the metamorphoses of nature. Never one of the most skilful among them could have imagined that it was destined to become a butterfly. It is the same with us. What more do we know of our destiny than of our origin? Let us then submit to an invincible ignorance on which our happiness depends.

He who so thinks will be wise, just, tranquil about his fate, and therefore happy. He will await death without either fear or desire, and will cherish life (hardly understanding how disgust can corrupt a heart in this place of many delights); he will be filled with reverence, gratitude, affection, and tenderness for nature, in proportion to his feeling of the benefits he has received from nature; he will be happy, in short, in feeling nature, and in being present at the enchanting spectacle of the universe, and he will surely never destroy nature either in himself or in others. More than that! Full of humanity, this man will love human character even in his enemies. Judge how he will treat others. He will pity the wicked without hating them; in his eyes, they will be but mis-made men. But in pardoning the faults of the structure of mind and body, he will none the less admire the beauties and the virtues of both. Those whom nature shall have favored will seem to him to deserve more respect than those whom she has treated in stepmotherly fashion. Thus, as we have seen, natural gifts, the source of all acquirements, gain from the lips and heart of the materialist, the homage which every other thinker unjustly refuses them. In short, the materialist, convinced, in spite of the protests of his vanity, that he is but a machine or an animal, will not maltreat his kind, for he will know too well the nature of those actions, whose humanity is always in proportion to the degree of the analogy proved above [between human beings and animals]; and following the natural law given to all animals, he will not wish to do to others what he would not wish them to do to him.

Let us then conclude boldly that man is a machine, and that in the whole universe there is but a single substance differently modified. This is no hypothesis set forth by dint of a number of postulates and assumptions; it is not the work of prejudice, nor even of my reason alone; I should have disdained a guide which I think to be so untrustworthy, had not my senses, bearing a torch, so to speak, induced me to follow reason by lighting the way themselves. Experience has thus spoken to me in behalf of reason; and in this way I have combined the two.

But it must have been noticed that I have not allowed myself even the most vigorous and immediately deduced reasoning, except as a result of a multitude of observations which no scholar will contest; and furthermore, I recognize only scholars as judges of the conclusions which I draw from the observations; and I hereby challenge every prejudiced man who is neither anatomist, nor acquainted with the only philosophy which can here be considered, that of the human body. Against so strong and solid an oak, what could the weak reeds of theology, of metaphysics, and of the schools, avail—childish arms, like our parlor foils, that may well afford the pleasure of fencing, but can never wound an adversary. Need I say that I refer to the empty and trivial notions, to the pitiable and trite arguments that will be urged (as long as the shadow of prejudice or of superstition remains on earth) for the supposed incompatibility of two substances which meet and move each other unceasingly? Such is my system, or rather the truth, unless I am much deceived. It is short and simple. Dispute it now who will.

[1]Haller in the *Transact. Philosoph.*

Hans Driesch

Hans Adolph Eduard Driesch was a German biologist born in Bad Kreuzach, Germany, in 1867. He studied zoology in Freiburg, Jena, and Munich. Driesch received his doctorate in Jena in 1889. He was an independent scholar in Heidelberg until 1920. From 1907 to 1908 he delivered the Gifford lectures in Edinburgh and in 1911 he became a professor of philosophy at Heidelberg. Subsequently, he taught in China, the United States, and Buenos Aires. He held a sharply antagonistic attitude toward Darwinism and found a great deal of sympathy with the biological views of Aristotle. Driesch became increasingly speculative about biological problems and his own experiments in embryogenesis. Ultimately he assumed an amphibious position between metaphysics and experimental biology and became professor of philosophy at Leipzig. Among his works are: *The Philosophy of Organism, The Crisis in Philosophy, The Problem of Individuality,* and *The History and Theory of Vitalism.* Driesch died in 1941.

The following selection is taken from *The Problem of Individuality,* pp. 1-19, published by The Macmillan Company in 1914.

THE PROBLEM OF INDIVIDUALITY

First Lecture

Introduction—Experimental Embryology—First Proof of Vitalism

Every problem of the philosophy of nature or, what is almost the same thing, every problem of theoretical science may be discussed in two very different ways. We may begin with what is generally called "the facts," or we may begin with the Ego as conceiving "facts"; we may either ascend or we may descend. In the first case we arrive at a certain

36

logical scheme postulated by the facts as they are, in the second we end by realizing that the facts discussed are the factual illustration of certain *a priori* possibilities. Neither of these two methods is strictly exclusive of the other, for, on the one hand, there is a good deal of logic in what is called "facts," and, on the other hand, there is something factual, so to speak, in all the concepts of logic, except the principle of identity, and especially in the general concept of *Nature*. But, nevertheless, the two methods may clearly be distinguished in practice, and this in every case, whether a problem of mechanics be the subject or a problem of biology.

There can be no doubt that the descending way is preferable from the philosophical point of view. We should start from the concepts of general logic as the *theory of order*, from concepts such as *this, such, relation, other, implication, member, arrangement, manifold*. We should develop the concept of *Nature* on the basis of everyday knowledge, and we should try then to discover what Reason makes of this strange thing called "Nature," *i.e.* what the general logical scheme of "Nature" *might possibly* be. At the end of all would come "the facts" of empirical science and would fit into certain places in the general logical scheme of "Nature," or even cover the whole.

But this kind of argument, though certainly superior to the other philosophically, because it is founded on the very essence of reason, is also much more difficult, at least for all who are not trained philosophers. All men are reasoning beings, but they do not *consciously* reason. We, in fact, do not realize what we are doing, what an enormously complicated logical operation we are performing, when we merely pronounce such an everyday phrase as "I got a letter from my friend this morning."

This being so, I shall not adopt the descending or deductive method of discussion alone in this course of lectures. Nor shall I pursue the ascending or empirically inductive method quite exclusively. To do so would detract from the absolute intrinsic necessity, the legitimate character, if I may say so, of the most general statements to be reached. I shall adopt *both* methods, one after the other, and let them come together. Let us, then, begin by formulating a certain well-known problem of natural science proper, without any particular philosophical aims at the outset.

Nobody will deny that the individual organism is of the type of a manifoldness which is at the same time a unity, that it represents a factual *wholeness*, if we may express its most essential character in a single technical word. And there is also not the least doubt that a great many of the processes occurring in the organism bring about this *wholeness*, or restore it if it is disturbed in any way. Processes of the first class are generally called *embryological* or ontogenetical. The restoring ones are spoken of as *restitution* or "regeneration" if the wholeness of the form as such is restored; they are described as *adaptation* if the physiological state of the organism has been disturbed and has now to be repaired; the factual wholeness represented by the organism being not merely a wholeness of form as such, but of living and functioning form. All of you know something, at least in rough outline, of the embryology of the frog; you have heard of the regeneration of the leg of a newt, and of the strange fact that in man one of the kidneys becomes larger if the other has been removed by an operation rendered necessary by some disease. These are three examples of processes which *bring about* or *restore* wholeness.

Let us now call all processes leading to factual wholeness *teleological* or purposeful processes. The expression "teleological" is for the moment to be nothing but a mere word, descriptive of a certain factual feature on the analogy of human acting. *There are* the individual organisms, each of them representing manifoldness in unity, *i.e.* factual wholeness, and *there are* processes, of at least three different kinds, embryological, restitutive, and adaptive, leading to this wholeness *as if* the existence of this wholeness were their "purpose." They always lead to wholeness; they have done so, and do so, and will do so, in innumerable cases.

So far there is a simple statement of fact, described by a certain technical name; there is as yet no problem. But a problem, in fact, *the* problem of biology at once arises, as soon as we consider a certain possibility that is suggested to us by another well-known fact. We are familiar with certain products of human workmanship which, factual wholenesses in themselves, produce other wholenesses by the processes which occur in them. These products of human work may, then, also be said to act "purposefully"; they are called *machines*; at least machines of certain kinds are of this type. Now, all single acts of becoming in a machine, taken by themselves, are of the *physico-chemical*, or mechanical, or, so to say, "inorganic" type. Wholeness, then, *may* be produced by a constallation of single *inorganic* or *mechanical* processes, in short by the working of a *machine*, and thus we are faced by the fundamental problem:

Is organic individual wholeness produced on the basis of a machine, i.e. *by processes which, though arranged in a special given manner, are*

*in themselves inorganic processes, as known from physics and chemis-
try, or are there in the organism whole-making processes* sui generis,
i.e. *processes not reducible to the forms of inorganic becoming?*

This, then, is the central problem of biology proper: *Mechanism or
Vitalism?* if by "Vitalism" we mean the possibility, merely negative at
first, that there *may* be processes in the organism which are *not* of the
machine-like or "mechanistic" type, and which may be said to be
"teleological" or purposeful in more than a merely formal sense.

It follows from the negative character which the concept of "vitalism"
must necessarily have at the outset, that the argument employed in
dealing with the great question must be of a particular logical type. If
ever we are able to "prove" vitalism, the proof can only be an apagogical
proof, or a proof *per exclusionem, i.e.* it can consist only in our becoming convinced that a machine *cannot* be the foundation of life. For the
concept of a *machine* is all that has been established as something
positive, so far; and the question is whether there be a machine *or not.*

It would be impossible in the course of these lectures, in which
biology proper forms only part of the subject, to discuss all classes of
biological phenomena at full length, and to inquire with respect to each
class whether "telology" is here of the machine-like or of some other
type. This I have done elsewhere, and I may be allowed to refer to my
published work on the subject.[1] Before my present audience I shall
select those biological facts that seem to me to be best suited to decide
our question, and shall mention the rest only in a few words. For it is not
with biology alone that we have to do in this course; biology is only to
yield us the solid foundation on which a factual—and not merely a
formal—understanding of the *universe* is to be obtained.

The facts of *active adaptation*—I do not speak here of "adaptedness,"
i.e. of being adapted, as a state—the facts of adaptation are very
numerous. Take for example what is called *functional adaptation, i.e.*
the fact that glands, muscles, bones and other tissues of the body
arrange their quantity and even their structure in correspondence to
changes of the general functional state, so that a bone, for instance, may
even adapt itself histologically to its being broken. Let me further
remind you of the adaptive structures of amphibious plants, adapted to
the water as well as to the air, and of the remarkable histological
adaptations of the larvae of *Salamandra*, according as they are reared in
the water or in the air, as discovered by Kammerer. There are also well-
known cases of purely physiological adaptation, unaccompanied by
histological changes; the regulation of heat-production in warm-blooded

animals belongs here, as does the selection of food materials out of given mixtures of food by Fungi, as discovered by Pfeffer—to mention only some of the most remarkable cases. Lastly, there is the production of so-called "antibodies" in correspondence to poisons and other substances, a fact which underlies the phenomena of immunity. In this case the range of active adaptation is very great, for the organism, at least of the higher vertebrates, is able to protect itself against an enormous variety of substances by the production of a material that counteracts their harmful effects.

This short survey has reminded you of well-known facts. What is the importance of these facts with regard to our central problem?

There cannot be the least doubt that all facts of adaptation are *teleological* in the sense defined; they re-establish functional wholeness after it has been disturbed; and we know that the organism is not only *whole* as regards its mere form, but that it is *whole* as a living, *i.e.* functioning form.

But, strange to say, none of the facts of adaptation, not even the curious facts of immunity and the production of "antibodies" have any decisive bearing on the question "mechanism or vitalism." Not that they are against vitalism in any way, if you are inclined to accept it; but they simply do not *prove* anything with regard to vitalism as the *only* possible form of a theory of life—and that is what a real theory of vitalism would require.

We have contrasted vitalism with the *machine*-theory of life. Now nobody could say from the facts of adaptation *taken by themselves* that a machine could *not* be the pre-established foundation of their happening. Such a machine would be very wonderful, very improbable even *qua* machine, in particular in the case of the production of antibodies to react against materials that had *never* entered the organism before. But the machine would not be *impossible*, and its *impossibility* must be demonstrated in order to establish vitalism.

Here, then, we may leave the facts of adaptation,[2] not without a certain feeling of disappointment; they can teach us nothing but what we are taught, say, by the selective faculties of the kidney. They may *in principle* be explained "mechanically," just as possible with respect to secretion, if only we attribute to the secreting organ, the kidney for example, a very complicated pre-established arrangement of its minute structure.

The study of adaptation, then, only teaches us a good deal with regard to the purposefulness of organic phenomena, but nothing more. Will the result be any more fertile, if we study the wonderful facts of *regeneration* in the same way? Strange to say, it will not. Regeneration in all its forms, be it regeneration of the embryo or of the adult, if only taken *as regeneration, i.e.* as a process repairing disturbed *wholeness*, would again make us familiar with a certain class of teleological processes, but would not do more. We should be dealing only with probabilities as regards the problem of vitalism.

But we want more; and we *can* gain more, if we only change our method of analysis. We must not attack *teleology* so directly and immediately in order to see whether it is of the machine-like or vitalistic type. We must devote ourselves to the facts without bias of any kind. It will be found that we get to real vitalism if we leave "teleology," at first, quite alone.

At the end of the 'eighties of the last century Professor Roux of Halle laid the foundations of *Entwicklungsmechanik*, a "new branch of anatomical science," as he called it. By the word *Entwicklungsmechanik* Roux means a branch of biology which may properly be called the *physiology of morphogenesis* or, in short, the *physiology of form*. It is an analytical and experimental science, just like physiology proper; ontogenesis, all kinds of restitution, heredity, phylogeny are its subjects.

Let us now enter a little more deeply into the *embryological* part of *Entwicklungsmechanik*.[3] Roux has worked out a sort of programme for this branch of the subject, and to it his own experimental investigations relate. At the beginning of his studies Roux was an *evolutionist*, almost in the same form as Weismann; and so-called evolutionism in embryology has always been a special form of machine-theory from the time of Leibniz to the present day. There is a very complicated machine in the egg and in particular in its nucleus—so Roux and Weismann said—and the development of the embryo is carried out by the *disintegration* of this machine during the great number of cell-cleavages which occur during the embryological process.

This was a possible theory, no doubt, and it seemed for a short time to be the right theory, for Roux happened to perform an experiment which, standing alone as it did, could really be considered as a sort of proof of embryological evolutionism.

Roux killed one of the first two cleavage cells of a frog's egg that had just performed the first cleavage; and from the surviving cell he reared an embryo which was in all respects *one half* of a normal one, that is to say, either the right or the left side of it. Was not this a very convincing result? It seemed so, no doubt—but only for a few years.

In 1891 I repeated Roux's experiment by a somewhat different method on the egg of the common sea-urchin. And my result was just the reverse of what Roux's result had been: not one half of an embryo was reared out of the surviving cell, but a *complete* embryo of *half size*. And I also observed the development of *complete* embryos of smaller size when I made my experiments with the four-cell-stage instead of the two-cell-stage. I might destroy one or two or even three of the first four cleavage cells; in the latter case I got a *very* small embryo—but it was *complete* in its organization.

Before we proceed in our argument let us make ourselves familiar with two technical concepts; this will prove to be very useful for what is to come. I mean the two concepts of *prospective value* and prospective potency, now quite familiar to embryological experimenters. By the *prospective value* of any embryonic cell whatever, I mean the *actual* fate which that cell will have in the special course of development going on before our eyes, be it normal or abnormal. By *prospective potency* I mean not the actual but the *possible* fate of a certain cell, *i.e.* the totality of possible characters of the adult into which this cell may develop.

Using these two concepts just defined, we may formulate what we have learned so far about the theories of Weismann and Roux and about the experimental results, in the following way. Roux and Weismann believed at first that the prospective value of a cleavage cell under normal conditions was *identical* with its prospective potency or, in other words, that its potency was strictly limited, and Roux believed he had proved this by his experiment with the frog's egg. But I was able to show that, for the egg of the sea-urchin at least, prospective value and prospective potency are *not the same*, the range of the prospective potency, *i.e.* the range of possibilities with regard to the morphogenetic fate, being *far greater* than the observation of the prospective value, *i.e.* of the actual fate in the actual case before me, could reveal.

I must next mention another experiment on the egg of the sea-urchin which is logically connected with what we have already learned.

The so-called "cleavage" of the egg, the first stages of which we have already considered, ends in the formation of the *blastula, i.e.* a hollow sphere built up of about a thousand cells, forming an epithelium. If you cut this blastula with a pair of very fine scissors in any direction you like, each part so obtained will go on developing—provided it is not smaller than one quarter of the whole—and will form a *complete* larva of small size. This result, certainly, might be expected after what we have learned from the experiment with the cleavage cells.

We are now at the right point for a theoretical discussion of our results. But before entering into it let us still devote a few words to the results of experiments carried out with eggs other than those of the frog and the sea-urchin. It has been shown that the eggs of very different classes of animals behave exactly as the egg of the sea-urchin does— namely, the eggs of Fishes, Newts, Amphioxus, Nemerteans, Medusae, etc. It has moreover been proved that even the frog's egg, the classical object of Roux's researches, produces a *small but complete* embryo from one of its cleavage cells, if only you give the cell an opportunity for a certain rearrangement of its protoplasm. And, finally, it is now known that in cases where, contrary to the behaviour of the Echinoderms, the prospective value of cleavage cells is truly fixed—as is the case in Annelids, Molluscs, and, to a certain extent, Ascidians—the fixation depends *solely* upon a certain physical state of the protoplasm, which does not allow of *any* regulatory rearrangement. It has been shown that in the forms with a fixed prospective value of the cleavage cells the nuclei, quite contrary to the theory of Weismann, are without any diversity, and that moreover there is no prospective specification in the protoplasm *before* cleavage really begins, or rather, to state it quite exactly, before so-called maturation. For you may alter in a fundamental manner the relative position of the nuclei of the cleavage cells with respect to one another by pressure experiments, or you may remove any portion you like from the egg before maturation: in both cases you will get complete embryos. Thus, then, our experimental results may be said to be of *universal* validity.

And now let us turn to the theoretical aspect.

How are we to account for what we have learned? A theory like Weismann's is impossible in the face of the facts. There is certainly *not* a machine in the egg that is disintegrated step by step during the cleavage, for *single* cleavage cells give *complete* organisms; and this relates to the protoplasm as well as to the nuclei. Might not, however, some other form of the machine theory fit the case?

In order to come to a conclusion in this difficult question I propose to formulate analytically, in quite a simple and unbiassed way, what our experiments have really shown us; and in particular I refer to the experiment with the blastula of the sea-urchin or the starfish.

Fragments of this blastula always gave complete embryos, though cut quite at random. This could only be possible, if the prospective potency, of all the thousand blastula cells was the same, just as the potency of the two or four first cleavage cells proved to be identical. Let us apply the term *equipotential ontogenetic system* to any ontogenetic totality which consists of cells with equal prospective potency, *i.e.* with an equal possible fate; then the blastula is, in short, an *equipotential system*.

But we must analyse our case still further, for there exist "equipotential" systems, which are very different from the blastula with regard to morphogenetic significance, in spite of their equipotentiality. The *ovary*, for instance, is certainly "equipotential," for each egg is "equally" able to form the organism; and yet there is a great logical difference between the ovary and the blastula. In the ovary each element of the system is equally able *to form for itself the same complex totality*, namely, the organism; we may speak of a "complex-equipotential system" in this case. But in the blastula each element is equally able *to play any single part in the formation of one totality*. Any particular cell would have played another single part, had you cut the blastula in some other direction; it can play *any* single part required. And what it actually does in the special case—normal or experimental—is always in *harmony* with what is done by its fellow-cells, which possess the same great potentiality as itself. Let us, then, call our blastula an harmonious-equipotential system.

On the discussion of the harmonious-equipotential system and its differentiation will depend our most important argument in favour of a vitalistic conception of biology. It is important, therefore, that this concept should become a little more familiar to you, and for this purpose the analysis of some other instances of harmonious equipotentiality is of great use. Harmonious systems not only appear elsewhere in embryology—the two so-called germ-layers, for instance, are of this type[4]—but very often they are the basis of *restitution*, which in this case is not "regeneration" proper, *i.e.* not a process of budding from a wound as is the case in the restitution of an earthworm cut into parts. The hydroid *Tubularia* offers a very typical instance of harmonious restitution; but more instructive still is the case of the restitution of the branchial apparatus in the Ascidian *Clavellina*, which therefore may be shortly analysed. In Clavellina the branchial apparatus is quite separated from the rest of the body. If you isolate it by a cut, it either regenerates the body in the usual way by budding processes, or it behaves very differently: it undergoes a complete reduction of form, until it is but a minute sphere, and then, after a few days of rest, transforms itself as it is into a *complete* little Ascidian. This, certainly, is a very strange process; but much more remarkable with regard to our problem is what follows. Isolate the branchial apparatus and then cut it into two pieces of any shape you like; each *portion* will then reduce its form, rest for a few days, and finally transform itself into a *complete* little animal, as did the whole branchial apparatus in the former experiment. The branchial apparatus of Clavellina, therefore, is the very type of a harmonious-equipotential system: each element of it is able to perform any single morphogenetic action that is required, and all the elements together work in harmony in each single case. For the cut may be made quite at random.

How, then, are we to account for these wonderful phenomena of life? Let us first enumerate all the possibilities of becoming that might seem to be present here at the first glance, but are found not to be present when you look at what happens in detail.

The question is this: *What makes the equipotential system unequal with regard to the actual fate of its parts*? What transforms equal potentialities into unequal actualities? In other words: the *localization* of the various singularities of morphogenesis is the problem. Whence does this localization come?

It does *not* come *from without*, for there are no localized exterior stimuli, responsible for differentiation in our cases of morphogenesis. The various factors or agents of the medium are either without direction or, if possessed of direction (*e.g.* gravity and light), they are notoriously without influence in animal ontogeny.

But localization can also *not* be based upon *purely chemical processes* inside the system. It is true, a chemical compound might be disintegrated, a real mixture might be separated into its components parts and the one or the other process *might a priori* be the main factor in ontogeny. But it cannot be so in fact. For from chemical disintegration or from unmixing there can only arise equilibria of, so to say, geometrical arrangement. But an organism is not a geometrical arrangement or a complex of such arrangements. And, further, there are many organs in an organism which have very different specific forms, though they have the same chemical composition—as for example the bones of vertebrates. For all this a *purely* chemical theory of ontogeny—which otherwise might be compatible with equipotentiality—cannot account.

But if a purely chemical theory of ontogenesis fails, might not some form of the *machine* theory be successful? Not, of course, the theory of Weismann, *i.e.* a theory of evolution or preformation in the narrow meaning of the word; but a theory which, nevertheless, makes use of the concept of a *machine* as the basis of ontogeny, a *machine* being defined as a *given specific combination of specific chemical and physical agents*. Ontogeny might then probably be the result of what would be called the "interaction" of these agents. Thus we know, for example, that the lens of the eye of Amphibians is formed from the epidermis in consequence of a so-called "formative stimulus" on the part of the primary optic vesicle; and there are other cases of morphogenesis of a similar kind.[5]

Now it is not difficult to prove from what we have learned about our harmonious-equipotential systems that *no* machine of any kind soever can be the ultimate basis of ontogenesis as far as harmonious equipotentiality is concerned.

If normal undisturbed embryogenesis alone would result in the formation of a complete embryo, if, in other words, all the experiments carried out with early embryonic stages would result in the production of fragments of organization, then we should feel obliged to accept the theory of machine-like preformation. But this is not the case. On the contrary, the ontogenetic systems are "harmonious-equipotential." Take whatever portion of them you like, quite at random, and *yet* there will be completeness of final organization. The embryonic "machine," then, that is supposed to exist in the normal system, would be obliged to be present in its completeness in one *part* of the system also, and also in another such part, and in yet other such parts too, and equally well in parts of different size, overlapping one another (Fig. 1). For we know that *any* part of the system, contingent as to its size and as to its position in the original system, can give rise to a complete being. *Every* cell of the original system can play *every* single role in morphogenesis; *which* role it will play is merely "a function of its position."

In face of *these* facts the machine theory as an embryological theory becomes an absurdity. These facts contradict the *concept* of a machine; for a machine is a specific arrangement of parts, and it does not remain what it was if you remove from it any portion you like.

Now the machine theory was the *only* possible form of a mechanistic theory that might *a priori* seem to be applicable to the phenomena of morphogenesis. To dismiss the machine theory, therefore, is the same as to give up the attempt of a mechanical theory of these phenomena altogether. Or, in other words, the analytical discussion of the differentiation of harmonious-equipotential systems entitles us to establish the doctrine of the *autonomy of life, i.e.* the doctrine of so-called vitalism, at least in a limited field: there is some agent at work in morphogenesis which is not of the type of physico-chemical agents.

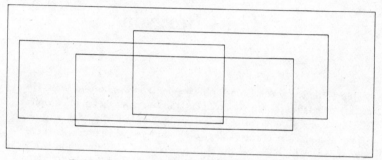

Fig. 1. The harmonious-equipotential system (H.E.S.).

The large rectangle represents an H.E.S. in its normal undisturbed state. It might *a priori* contain a very complicated kind of "machine" as the foundation of development. But any fragment of the system (the small rectangles and innumerable others), contingent as to its size and to its position in the original H.E.S., is equally able to produce a small *but complete* organism. On the basis of the mechanistic theory, then, any fragment of the H.E.S. would contain the same "machine" as the original system. This is absurd.

NOTES

[1] *The Science and Philosophy of the Organism*, The Gifford Lectures delivered before the University of Aberdeen in the years 1907-8, 2 vols. London, Adam & Charles Black, 1908.

[2] For a full discussion of the facts of adaptation refer to *Gifford Lectures*, vol. i. pp. 165-213.

[3] Cp. *Gifford Lectures*, vol. i. pp. 25-164.

[4] A very important case of harmonious equipotentiality, not mentioned in my *Gifford Lectures*, was afterwards discovered by Braus (see *Morphologisches Jahrbuch*, vol. 39).

[5] Cp. Herbst, *Biologisches Centralblatt*, vol. xiv. and xv. (1895-4), and *Formative Reise in der tierischen Ontogenese*, Leipzig, 1902.

[6] See, on this subject, *Matiere et memoire*, chap. i.

Henri Bergson

Henri Bergson was a French philosopher who was born in 1859. He was educated at the University of Paris and was a teacher of philosophy at the Lycee Condorcet in Paris from 1889 to 1897. In 1900 Bergson joined the faculty of the College de France and was a professor there until 1921. He delivered the Gifford lectures in Edinburgh in 1912. The Nobel prize for literature was awarded him in 1927. His philosophy became something of a rage for a time in Europe, but then declined in popularity before his death. Some of his works are: *Introduction to Metaphysics*, *Two Sources of Morality and Religion*, and *Creative Evolution*. Bergson died in 1941.

The following selection is taken from *Creative Evolution*, translated by Arthur Mitchell in the Modern Library Edition, pp. 97-108. The first American edition was published by Henry Holt and Company in 1911.

CREATIVE EVOLUTION

So we come back, by a somewhat roundabout way, to the idea we started from, that of an *original impetus* of life, passing from one generation of germs to the following generation of germs through the developed organisms which bridge the interval between the generations. This impetus, sustained right along the lines of evolution among which it gets divided, is the fundamental cause of variations, at least of those that are regularly passed on, that accumulate and create new species. In general, when species have begun to diverge from a common stock, they accentuate their divergence as they progress in their evolution. Yet, in certain definite points, they may evolve identically; in fact, they must do so if the hypothesis of a common impetus be accepted. This is just what we shall have to show now in a more precise way, by the same example we have chosen, the formation of the eye in mollusks and vertebrates. The idea of an "original impetus," moreover, will thus be made clearer.

Two points are equally striking in an organ like the eye: the complexity of its structure and the simplicity of its function. The eye is composed of distinct parts, such as the sclerotic, the cornea, the retina, the crystalline lens, etc. In each of these parts the detail is infinite. The retina alone comprises three layers of nervous elements—multipolar cells, bipolar cells, visual cells—each of which has its individuality and is undoubtedly a very complicated organism: so complicated, indeed, is the retinal membrane in its intimate structure, that no simple description can given an adequate idea of it. The mechanism of the eye is, in short, composed of an infinity of mechanisms, all of extreme complexity. Yet vision is one simple fact. As soon as the eye opens, the visual act is effected. Just because the act is simple, the slightest negligence on the part of nature in the building of the infinitely complex machine would have made vision impossible. This contrast between the complexity of the organ and the unity of the function is what gives us pause.

A mechanistic theory is one which means to show us the gradual building-up of the machine under the influence of external circumstances intervening either directly by action on the tissues or indirectly by the selection of better-adapted ones. But, whatever form this theory may take, supposing it avails at all to explain the detail of the parts, it throws no light on their correlation.

Then comes the doctrine of finality, which says that the parts have been brought together on a preconceived plan with a view to a certain end. In this it likens the labor of nature to that of the workman, who also proceeds by the assemblage of parts with a view to the realization of an idea or the imitation of a model. Mechanism, here, reproaches finalism with its anthropomorphic character, and rightly. But it fails to see that itself proceeds according to this method—somewhat mutilated! True, it has got rid of the end pursued or the ideal model. But it also holds that nature has worked like a human being by bringing parts together, while a mere glance at the development of an embryo shows that life goes to work in a very different way. *Life does not proceed by the association and addition of elements, but by dissociation and division.*

We must get beyond both points of view, both mechanism and finalism being, at bottom, only standpoints to which the human mind has been led by considering the work of man. But in what direction can we go beyond them? We have said that in analyzing the structure of an organ, we can go on decomposing forever, although the function of the whole is a simple thing. This contrast between the infinite complexity of the organ and the extreme simplicity of the function is what should open our eyes.

In general, when the same object appears in one aspect as simple and in another as infinitely complex, the two aspects have by no means the same importance, or rather the same degree of reality. In such cases, the simplicity belongs to the object itself, and the infinite complexity to the views we take in turning around it, to the symbols by which our senses or intellect represent it to us, or, more generally, to elements *of a different order*, with which we try to imitate it artificially, but with which it remains incommensurable, being of a different nature. An artist of genius has painted a figure on his canvas. We can imitate his picture with many-colored squares of mosaic. And we shall reproduce the curves and squares of the model so much the better as our squares are smaller, more numerous and more varied in tone. But an infinity of elements infinitely small, presenting an infinity of shades, would be necessary to obtain the exact equivalent of the figure that the artist has conceived as a simple thing, which he has wished to transport as a whole to the canvas, and which is the more complete the more it strikes us as the projection of an indivisible intuition. Now, suppose our eyes so made that they cannot help seeing in the work of the master a mosaic effect. Or suppose our intellect so made that it cannot explain the appearance of the figure on the canvas except as a work of mosaic. We should then be able to speak simply of a collection of little squares, and we should be under the mechanistic hypothesis. We might add that, besides the materiality of the collection, there must be a plan on which the artist worked; and then we should be expressing ourselves as final-ists. But in neither case should we have got at the real process, for there are no squares brought together. It is the picture, *i.e.* the simple act, projected on the canvas, which, by the mere fact of entering into our perception, is *de*composed before our eyes into thousands and thousands of little squares which present, as *re*composed, a wonderful arrangement. So the eye, with its marvelous complexity of structure, may be only the simple act of vision, divided *for us* into a mosaic of cells, whose order seems marvelous to us because we have conceived the whole as an assemblage.

If I raise my hand from A to B, this movement appears to me under two aspects at once. Felt from within, it is a simple, indivisible act. Perceived from without, it is the course of a certain curve, AB. In this curve I can distinguish as many positions as I please, and the line itself might be defined as a certain mutual co-ordination of these positions. But the positions, infinite in number, and the order in which they are connected, have sprung automatically from the indivisible act by which my hand has gone from A to B. Mechanism, here, would consist in seeing only the positions. Finalism would take their order into account. But both mechanism and finalism would leave on one side the movement, which is reality itself. In one sense, the movement is *more* than the positions and than their order; for it is sufficient to make it in its indivisible simplicity to secure that the infinity of the successive positions as also their order be given at once—with something else which is neither order nor position but which is essential, the mobility. But, in another sense, the movement is *less* than the series of positions and their connecting order; for, to arrange points in a certain order, it is necessary first to conceive the order and then to realize it with points; there must be the work of assemblage and there must be intelligence, whereas the simple movement of the hand contains nothing of either. It is not intelligent, in the human sense of the word, and it is not an assemblage, for it is not made up of elements. Just so with the relation of the eye to vision. There is in vision *more* than the component cells of the eye and their mutual co-ordination: in this sense, neither mechanism nor finalism go far enough. But, in another sense, mechanism and finalism both go too far, for they attribute to Nature the most formidable of the labors of Hercules in holding that she has exalted to the simple act of vision an infinity of infinitely complex elements, whereas Nature has had no more trouble in making an eye than I have in lifting my hand. Nature's simple act has divided itself automatically into an infinity of elements which are then found to be co-ordinated to one idea, just as the movement of my hand has dropped an infinity of points which are then found to satisfy one equation.

We find it very hard to see things in that light, because we cannot help conceiving organization as manufacturing. But it is one thing to manufacture, and quite another to organize. Manufacturing is peculiar to man. It consists in assembling parts of matter which we have cut out in such manner that we can fit them together and obtain from them a common action. The parts are arranged, so to speak, around the action as an ideal center. To manufacture, therefore, is to work from the periphery to the center, or, as the philosophers say, from the many to the one. Organization, on the contrary, works from the center to the periphery. It begins in a point that is almost a mathematical point, and spreads around this point by concentric waves which go on enlarging.

The work of manufacturing is the more effective, the greater the quantity of matter dealt with. It proceeds by concentration and compression. The organizing act, on the contrary, has something explosive about it: it needs at the beginning the smallest possible place, a minimum of matter, as if the organizing forces only entered space reluctantly. The spermatozoon, which sets in motion the evolutionary process of the embryonic life, is one of the smallest cells of the organism; and it is only a small part of the spermatozoon which really takes part in the operation.

But these are only superficial differences. Digging beneath them, we think, a deeper difference would be found.

A manufactured thing delineates exactly the form of the work of manufacturing it. I mean that the manufacturer finds in his product exactly what he has put into it. If he is going to make a machine, he cuts out its pieces one by one and then puts them together: the machine, when made, will show both the pieces and their assemblage. The whole of the result represents the whole of the work; and to each part of the work corresponds a part of the result.

Now I recognize that positive science can and should proceed as if organization was like making a machine. Only so will it have any hold on organized bodies. For its object is not to show us the essence of things, but to furnish us with the best means of acting on them. Physics and chemistry are well advanced sciences, and living matter lends itself to our action only so far as we can treat it by the processes of our physics and chemistry. Organization can therefore only be studied scientifically if the organized body has first been likened to a machine. The cells will be the pieces of the machine, the organism their assemblage, and the elementary labors which have organized the parts will be regarded as the real elements of the labor which has organized the whole. This is the standpoint of science. Quite different, in our opinion, is that of philosophy.

For us, the whole of an organized machine may, strictly speaking, represent the whole of the organizing work (this is, however, only approximately true), yet the parts of the machine do not correspond to parts of the work, because *the materiality of this machine does not represent a sum of means employed, but a sum of obstacles avoided*: it is a negation rather than a positive reality. So, as we have shown in a former study, vision is a power which should attain *by right* an infinity of things inaccessible to our eyes. But such a vision would not be continued into action; it might suit a phantom, but not a living being. The vision of a living being is an *effective* vision, limited to objects on which the being can act: it is a vision that is *canalized*, and the visual apparatus simply symbolizes the work of canalizing. Therefore the creation of the visual apparatus is no more explained by the assembling of its anatomic elements than the digging of a canal could be explained by the heaping-up of the earth which might have formed its banks. A mechanistic theory would maintain that the earth had been brought cart-load by cart-load; finalism would add that it had not been dumped down at random, that the carters had followed a plan. But both theories would be mistaken, for the canal has been made in another way.

With greater precision, we may compare the process by which nature constructs an eye to the simple act by which we raise the hand. But we supposed at first that the hand met with no resistance. Let us now imagine that, instead of moving in air, the hand has to pass through iron filings which are compressed and offer resistance to it in proportion as it goes forward. At a certain moment the hand will have exhausted its effort, and, at this very moment, the filings will be massed and coordinated in a certain definite form, to wit, that of the hand that is stopped and of a part of the arm. Now, suppose that the hand and arm are invisible. Lookers-on will seek the reason of the arrangement in the filings themselves and in forces within the mass. Some will account for the position of each filing by the action exerted upon it by the neighboring filings: these are the mechanists. Others will prefer to think that a plan of the whole has presided over the detail of these elementary actions: they are the finalists. But the truth is that there has been merely one indivisible act, that of the hand passing through the filings: the inexhaustible detail of the movement of the grains, as well as the order of their final arrangement, expresses negatively, in a way, this undivided movement, being the unitary form of a resistance, and not a synthesis of positive elementary actions. For this reason, if the arrangement of the grains is termed an "effect" and the movement of the hand a "cause," it may indeed be said that the whole of the effect is explained by the whole of the cause, but to parts of the cause parts of the effect will in no wise correspond. In other words, neither mechanism nor finalism will here be in place, and we must resort to an explanation of a different kind. Now, in the hypothesis we propose, the relation of vision to the visual apparatus would be very nearly that of the hand to the iron filings that follow, canalize and limit its motion.

The greater the effort of the hand, the farther it will go into the filings. But at whatever point it stops, instantaneously and automatically the filings co-ordinate and find their equilibrium. So with vision and its organ. According as the undivided act constituting vision advances more or less, the materiality of the organ is made of a more or less considerable number of mutually coordinated elements, but the order is necessarily complete and perfect. It could not be partial, because, once again, the real process which gives rise to it has no parts. That is what neither mechanism nor finalism takes into account, and it is what we also fail to consider when we wonder at the marvelous structure of an instrument such as the eye. At the bottom of our wondering is always this idea, that it would have been possible for *a part only* of this coordination to have been realized, that the complete realization is a kind of special favor. This favor the finalists consider as dispensed to them all at once, by the final cause; the mechanists claim to obtain it little by little, by the effect of natural selection; but both see something positive in this co-ordination, and consequently something fractionable in its cause—something which admits of every possible degree of achievement. In reality, the cause, though more or less intense, cannot produce its effect except in one piece, and completely finished. According as it goes further and further in the direction of vision, it gives the simple pigmentary masses of a lower organism, or the rudimentary eye of a Serpula; or the slightly differentiated eye of the Alciope, or the marvelously perfected eye of the bird; but all these organs, unequal as is their complexity, necessarily present an equal co-ordination. For this reason, no matter how distant two animal species may be from each other, if the progress toward vision has gone equally far in both, there is the same visual organ in each case, for the form of the organ only expresses the degree in which the exercise of the function has been obtained.

But, in speaking of a progress toward vision, are we not coming back to the old notion of finality? It would be so, undoubtedly, if this progress required the conscious or unconscious idea of an end to be attained. But it is really effected in virtue of the original impetus of life; it is implied in this movement itself, and that is just why it is found in independent lines of evolution. If now we are asked why and how it is implied therein, we reply that life is, more than anything else, a tendency to act on inert matter. The direction of this action is not predetermined; hence the unforeseeable variety of forms which life, in evolving, sows along its path. But this action always presents, to some extent, the character of contingency; it implies at least a rudiment of choice. Now a choice involves the anticipatory idea of several possible actions. Possibilities of action must therefore be marked out for the living being before the action itself. Visual perception is nothing else:* the visible outlines of bodies are the design of our eventual action on them. Vision will be found, therefore, in different degrees in the most diverse animals, and it will appear in the same complexity of structure wherever it has reached the same degree of intensity.

We have dwelt on these resemblances of structure in general, and on the example of the eye in particular, because we had to define our attitude toward mechanism on the one hand and finalism on the other. It remains for us to describe it more precisely in itself. This we shall now do by showing the divergent results of evolution not as presenting analogies, but as themselves mutually complementary.

Ludwig von Bertalanffy

Ludwig von Bertalanffy, who was professor of theoretical biology at the State University of New York at Buffalo, is perhaps best known as the founder of general systems theory. He was born in 1901 in Austria and received the Ph.D. at the University of Vienna in 1926. Bertalanffy served on the faculty there from 1934 to 1948. He was professor and director of biological research at the University of Ottawa from 1949 to 1954 and visiting professor at the University of Southern California at Los Angeles from 1955 to 1958. He is the author of numerous articles and books including: *Modern Theories of Development, Theoretische Biologie, Problems of Life*, and *General Systems Theory*. Bertalanffy died in Buffalo in 1972.

The following selection is from *Problems of Life* (1952) pp. 9-22, and is reprinted here with the kind permission of Harper & Row Publishers, Inc.

PROBLEMS OF LIFE

2. The Organismic Conception

In our time a fundamental change of scientific conceptions has occurred. The revolutions in modern physics are widely known. They have led, in the relativity and quantum theories, to a radical reform and expansion of physical doctrine, outranking the progress made in centuries of the past. Less obvious, but perhaps not less significant in their consequences, are the changes that have taken place in biological thought, changes that have led both to a new attitude towards the basic problems of living nature and to new questions and solutions.

We might take as an established fact of the modern development in biology that it does not consent completely to either of the classical views, but transcends both in a new and third one. This attitude has been called the *organismic conception* by the author, who has worked it out for more than twenty years. Similar conceptions have been found necessary, and have been evolved, in the most diverse fields of biology, as well as in the neighbouring sciences of medicine, psychology, sociology, etc. If we retain the term "organismic conception" we shall consider it merely as a convenient denomination for an attitude which has already become very general and largely anonymous. This seems to be justified, in so far as the author was probably the first to develop the new standpoint in a scientifically and logically consistent form.

Biological research and thought have hitherto been determined by three leading ideas, which may be called the *analytical and summative*, the *machine-theoretical*, and the *reaction-theoretical conceptions*.

It appeared to be the goal of biological research to resolve the complex entities and processes that confront us in living nature into elementary units—to *analyze* them—in order to explain them by means of the juxtaposition or *summation* of those elementary units and processes. Procedure in classical physics supplied the pattern. Thus chemistry resolves material bodies into elementary components—molecules and atoms; physics considers a storm that tears down a tree as the sum of movements of air particles, the heat of a body as the sum of the energy of motion of molecules, and so on. A corresponding procedure was applied in all biological fields, as some examples will easily show.

Thus biochemistry investigates the individual chemical constituents of living bodies and the chemical processes going on within them. In this way it specifies the chemical compounds found in the cell and the organism, as well as their reactions.

The classical cell theory considered cells as the elementary units of life, comparable to atoms as the elementary units of chemical compounds. So a multicellular organism appeared morphologically as an aggregate of such building units; physiologically, it was the tendency to resolve the processes in the whole organism into the processes within the cells. Virchow's "cellular pathology" and Verworn's "cellular physiology" gave a programmatic statement of this attitude.

The same point of view was applied to the embryonic development of organisms. Weismann's classical theory assumed that there exists in the egg nucleus a number of *anlagen* or tiny elementary developmental machines for building the individual organs. In the course of the development, these are progressively segregated by means of the cell divisions the germ is passing through, and thus located in different regions. They bestow on those regions their specific characters, and so finally determine the histological and anatomical structure in the fully developed organism.

Of great importance, not only from the theoretical but also from the clinical viewpoint, is the classical theory of reflexes, centres, and localization. The nervous system was considered to be a sum of apparatuses established for individual functions. For example, in the spinal cord segmental centres for the individual reflexes are present; similarly in the brain centres for the various fields of conscious sense-perceptions, for the voluntary movements of individual muscle-groups, for speech and the other higher mental activities. Accordingly, the behaviour of animals was resolved into a sum or chain of reflexes.

Genetics considered the organism as an aggregate of characters going back to a corresponding aggregate of genes in the germ cells, transmitted and acting independently of each other.

Accordingly, the theory of natural selection resolved living beings into a complex of characters, some useful, others disadvantageous, which characters, or rather their corresponding genes, are transmitted independently, thus through natural selection affording the opportunity for the elimination of unfavourable characters, while allowing favourable ones to survive and accumulate.

The same principle could be shown to operate in every field of biology, and in medicine, psychology, and sociology as well. The examples given will suffice, however, to show that the principle of analysis and summation has been directive in all fields.

Analysis of the individual parts and processes in living things is *necessary*, and is the prerequisite for all deeper understanding. Taken alone, however, analysis is not *sufficient*.

The phenomena of life—metabolism, irritability, reproduction, development, and so on—are found exclusively in natural bodies which are circumscribed in space and time, and show a more or less complicated structure; bodies that we call "organisms." Every organism represents a *system*, by which term we mean a complex of elements in mutual interaction.

From this obvious statement the limitations of the analytical and summative conceptions must follow. First, it is impossible to resolve the phenomena of life completely into elementary units; for each individual part and each individual event depends not only on conditions within itself, but also to a greater or lesser extent on the conditions within the *whole*, or within superordinate units of which it is a part. Hence the behaviour of an isolated part is, in general, different from its behaviour within the context of the whole. The action of an isolated blastomere in Driesch's experiment is different from what it is in the whole embryo. If cells are explanted from the organism and allowed to grow as a tissue culture in an appropriate nutrient, their behaviour will be different from that within the organism. The reflexes of an isolated part of the spinal cord are not the same as the performances of these parts in the intact nervous system. Many reflexes can be demonstrated clearly only in the isolated spinal cord, whereas in the intact animal the influence of higher centres and the brain alters them decidedly. Thus the characteristics of life are characteristics of a system arising from, and associated with, the organization of materials and processes. Thus they are altered with alterations in the whole, and disappear when it is destroyed.

Secondly, the actual whole shows properties that are absent from its isolated parts. The problem of life is that of *organization*. As long as we single out individual phenomena we do not discover any fundamental difference between the living and the non-living. Certainly organic molecules are more complicated than inorganic ones; but they are not distinguishable from dead compounds by fundamental differences. Even complicated processes, considered a long time as being specifically vital, like those of cell respiration and fermentation, morphogenesis, nerve action, and so on, have been explained to a large extent physico-chemically, and many of them can even be imitated in inanimate models. A fundamentally new problem is presented, however, in the singular and specific arrangement of parts and processes that we meet with in living systems. Even a knowledge of all the chemical compounds that build a cell would not explain the phenomena of life. Already the simplest cell is a superlatively complex organization, the laws of which are at present only dimly seen. A "living substance" has often been spoken of. This concept is due to a fundamental fallacy. There is no "living substance" in the sense that lead, water, or cellulose are substances, where any arbitrarily taken part shows the same properties as the rest. Rather is life bound to individualized and organized systems, the destruction of which puts an end to it.

Similar considerations apply to the processes of life. So long as we consider the individual chemical reactions that take place in a living organism we are unable to indicate any basic difference between them and those that go on in inanimate things or in a decaying corpse. But a fundamental contrast is found when we consider, not single processes, but their totality within an organism or a partial system of it, such as a cell or an organ. Then we find that all parts and processes are so ordered that they guarantee the maintenance, construction, restitution, and reproduction of organic systems. This order basically distinguishes events in a living organism from reactions taking place in non-living systems or in a corpse.

This has been depicted vividly as follows:

> Unstable substances dissociate; combustible ones burn occasionally; catalysers accelerate slow processes. There is nothing extraordinary about this. But that catabolism does not destroy the organism which it is continually nibbling, but on the contrary indirectly maintains it, makes it an organic process. That the constant glow in our tissues does not attack their structure; that therefore every animal and plant resembles a steam engine made of fuel and yet incessantly working; in this fact is respiration distinguished from ordinary oxidation. Just so excretion would be an osmotic phenomenon like any other if it were not for the fact that the glands remove what is noxious for the organism and retain what is valuable. We can easily explain the movements of plants and lower animals as reactions to stimuli; and who is willing to avoid a sharp dividing line across the animal kingdom, will lastly interpret also spontaneous movements in the same way; to him they are "brain reflexes," very complicated indeed, but not essentially different from simple reflexes that take place in reaction to external stimuli. Now let us imagine that a dead reflex-apparatus is constructed. It must be charged with latent energies; even slight disturbances would be able to release powerful movements; a special apparatus would provide for continual storage of potential energy. In what way would such mechanism differ fundamentally from a living being, the action on it from a stimulus, its movement from organic movement? In the fact that all organic reactions directly or indirectly serve to maintain existent, or to produce demanded forms. (J. Schultz, 1929)

Thus the problem of wholeness and organization sets a limit to the analytical and summative description and explanation. In what way is it accessible to scientific investigation?

Classical physics, the conceptual scheme of which was adopted in biology, was to a large extent summative in character. In mechanics it could consider a body as a sum, in heat theory, a gas as a chaos, of mutually independent molecules. In fact, the word "gas," introduced in the sixteenth century by the physician van Helmont, denoted just "chaos," in unconscious symbolism. In modern physics, however, the principles of wholeness and organization gain a hitherto unexpected significance. Atomic physics everywhere encounters wholes that cannot be resolved into the behaviour of elements considered in isolation. Whether atomic structure or structural formulae of chemical compounds or space-lattices of crystals are investigated, problems of organization always arise and appear to be the most urgent and fascinating of modern physics. From such viewpoint, the analytical and summative attitude towards the living seems to be a tremendous solecism. A silly dead crystal has a marvellous architectónic, the design of which makes the mathematical physicist's reasoning work at its utmost speed. But living protoplasm, with its astonishing properties, was thought to have been explained when it was called a "colloidal solution." An atom or a crystal are not the result of chance forces but of organizational ones; yet it was thought possible to explain the organized things *par excellence*, the living organisms, as chance products of mutation and selection.

The task of biology, therefore, is to establish the laws governing order and organization within the living. Moreover, as we shall see presently, these laws are to be investigated at all levels of biological organization—at the physico-chemical level, at the level of the cell and of the multicellular organization, and finally at the level of communities consisting of many individual organisms.

How is biological organization to be interpreted?

All knowledge starts from sensory experience. The primary tendency, therefore, is to devise visualizable models. When, for example, science came to the conclusion that elementary units called atoms are at the basis of reality, its first conception was that of tiny hard bodies similar to miniature billiard balls. Not until later was it realized that this is not so, and that the final units are entites not to be defined by visual models, but only by mathematical abstractions, concepts like "matter" and "energy," "corpuscle" and "wave" indicating only certain aspects of their behaviour. When mankind watched the spectacle of the regular movements of the stars, first they looked for powerful machineries, the rotation of which keeps the stars going in harmonic motion—those crystal spheres Aristotle dreamed of—until astonomy destroyed this picture, learning that the order of planetary movements is due only to the mutual attraction of the heavenly bodies in the empty space. Thus structure is the first thing the human mind looks for to explain the order of natural processes; an explanation in terms of organizing forces is much more difficult.

This applies also to the explanation of life. Observing the inconceivable multiplicity of processes going on in the cell or in the organism, in order to maintain its subsistence, only one explanation seemed possible. It is what may be called the *machine theory*, meaning that the order in vital phenomena was to be interpreted in terms of structures, mechanisms in the widest sense of the word. Examples of this conception are Weismann's theory of embryonic development, or the classic reflex and centre theory; but the same type of explanation can be found in every field of biology.

Now structural conditions are to a large extent present in the living organism. The physiology of organs—for example, of organs of nutrition, circulation, secretion, of sense organs as receptors for stimuli, of the nervous system and its connections, and so on—is nothing but a description of the technical masterpiece which confronts us in an organism. In the same way we find structures as mediators of order in every cell, from the muscle and nerve fibrils, as apparatus for contraction and the conduction of excitation, to the cell organs of secretion and division, the chromosomes as structural units of heredity, and so forth.

Nevertheless, we cannot consider structures as the primary basis of the vital order, for three reasons.

First, in all realms of living phenomena we find the possibility of regulation following disturbances. Driesch is right that such regulation, for example, in embryonic development, would be impossible on the basis of a "machine," for a fixed structure can respond to certain definite exigencies only, not just to any one whatever.

Secondly, there is a fundamental difference between the structure of a machine and that of an organism. The former consists always of the same components, the latter is maintained in a state of continuous flux, a perpetual breaking down and replacement of its building materials. Organic structures are themselves the expression of an ordered process, and are only maintained in and by this process. Therefore, the primary order of organic processes must be sought in the processes themselves, not in pre-established structures.

Thirdly, ontogenetically as well as phylogenetically we find a transition from less mechanized and more regulable states to more mechanized and less regulable ones. To illustrate this again by an example from embryonic development: If at an early stage a piece of presumptive epidermis of an amphibian embryo is transplanted to the region of the future brain, it becomes part of the brain. At a later stage, however, the embryonic regions are determined irrevocably to form certain organs. Thus a piece of presumptive brain will become, even after displacement, brain or a derivative, for example, an eye which develops in the coelomic cavity, and is here, of course, totally misplaced. A similar fixation to only one function, a progressive mechanization as we may call it, is found in the most diverse phenomena of life.

We come therefore to the following conclusion. Primarily, organic processes are determined by the mutual interaction of the conditions present in the total system, by a *dynamic* order as we may call it. This is at the basis of organic regulability. Secondarily, a progressive mechanization takes place, i.e., the originally unitary action segregates into separate actions, governed by fixed structures. The primary nature of dynamic as opposed to a structural or machine-like order, is seen in fields as diverse as those of cell structures, embryonic development, secretion, phagocytosis and resorption, the theory of reflexes and centres, of instinctive behaviour, *gestalt* perception, etc. Organisms *are not* machines, but they can to a certain extent *become* machines, congeal into machines. Never completely, however, for a thoroughly mechanized organism would be incapable of regulation following disturbances, or of reacting to the incessantly changing conditions of the outside world. The fact that organic processes never represent a mere sum of single structurally fixed processes, but to a greater or less extent always have the character of processes determined within a dynamic system, gives them adaptability to changing circumstances and regulability following disturbances.

The comparison of the organism with a machine also leads to the last of the points of view we have mentioned, the one we call the *reaction theory*. The organism was considered as a sort of automaton. Just as a penny-in-the-slot machine, by virtue of an internal mechanism, delivers an article after a coin has been inserted, so the organism responds to the stimulation of a sense organ with a certain reflex action, to the intake of food with the production of certain enzymes, and so forth. Thus, the organism was considered an essentially passive system, set into action only through outside influences, the so-called stimuli. This "stimulus-response scheme" has been of fundamental importance, especially in the theory of animal behaviour.

In fact, however, the organism is, even under constant external conditions and in the absence of external stimuli, not a passive but a basically *active* system. This is obvious in the fundamental phenomenon of life, metabolism, the continuous building up and breaking down of component materials, which is inherent in the organism and not forced upon it by external conditions. This viewpoint becomes especially important in considering the activity of the nervous system, irritability, and behaviour. Modern research has shown that we have to consider autonomous activity, as it is manifest, for example, in the rhythmic-automatic functions, as the primary phenomenon rather than reflexes and reactivity.

We can therefore summarize the leading principles of an organismic conception in the following way: *The conception of the system as a whole* as opposed to the *analytical* and *summative* points of view; the *dynamic conception* as opposed to the *static* and *machine-theoretical* conceptions; the consideration of the organism as a *primary activity* as opposed to the conception of its *primary reactivity*.

These principles enable us to overcome the antagonism of the mechanistic and vitalistic conceptions. Both are based on the analytical, summative, and machine-theoretical principles. The mechanistic theory did not approach just the fundamental problems of life—order, organization, wholeness, and self-regulation. These remained unsolved by analytical investigation, and the attempt to explain them by way of the machine theory, i.e., on the basis of pre-existing structures, leads to failure in dealing with basic phenomena and problems. Vitalism starts with these unsolved problems. But it does not overthrow the summative and machine-theoretical conceptions. On the contrary, vitalism views a living organism as a sum of parts and machine-like structures, assuming them to be controlled and supplemented by a soul-like engineer. Thus Driesch, for example, declared the embryo to be a "sum-like aggregate" of cells, converted into a whole by entelechy. Thus, instead of starting with an unbiased view of the organic system, vitalists also start with the preconceived conception of the organic machine. They realize that, in view of the phenomenon of regulation and of the origin of the machine, this conception is not satisfactory. In order to save it, they introduce factors that repair the machines after disturbance or act as their maker. Thus only two possible explanations of organic order and regulation have been recognized: orderliness through fixed machine-like structures, or as the result of some vitalistic factor. Both are inadequate. The mechanistic view breaks down in face of the phenomenon of regulation and of the origin of the "machine"; vitalism renounces scientific explanation.

Opposed to both, stands an organismic conception. For understanding life phenomena it is neither sufficient to know the individual elements and processes nor to interpret their order by means of machine-like structures, even less to invoke an entelechy as the organizing factor. It is not only necessary to carry out analysis in order to know as much as possible about the individual components, but it is equally necessary to know the laws of organization that unite these parts and partial processes and are just the characteristic of vital phenomena. Herein lies the essential and original object of biology. This biological order is specific and surpasses the laws applying in the inanimate world, but we can progressively approach it with continued research. It calls for investigation at all levels: at the level of physico-chemical units, processes, and systems; at the biological level of the cell and the multicellular organism; at the level of supra-individual units of life. At each of these levels we see new individual units of life. At each of these levels we see new properties and new laws. Biological order is, in a wide measure, of a dynamic nature; how this is to be defined we shall see later on.

In this way the autonomy of life, denied in the mechanistic conception, and remaining a metaphysical question mark in vitalism, appears, in the organismic conception, as a problem accessible to science and, in fact, already under investigation.

The term "wholeness" has been much misused in past years. Within the organismic conception it means neither a mysterious entity nor a refuge for our ignorance, but a fact that can and must be dealt with by scientific methods.

The organismic conception is not a compromise, a muddling through or mid-course between the mechanistic and vitalistic views. As we have seen, the analytic, summative, and machine-theoretical conceptions have been the common ground of both the classical views. Organization and wholeness considered as principles of order, immanent to organic systems, and accessible to scientific investigation, involve a basically new attitude. What occurred to the organismic conception was, however, what usually happens to new ideas: first it was attacked and refused, then declared to be old and self-evident. In fact, once it is realized, this conception merely draws the consequences from the obvious statement that organisms are organized. To achieve this unbiased approach it was necessary, however, and in many fields is still necessary, even today, to combat deeply rooted habits of thought.

The organismic conception must be examined, first, in its significance as a *method of research and theory in biology*; secondly, in its *epistemological significance*.

Busy with special questions and experiments, the research worker in the laboratory is looking at "general considerations" with mistrust and aversion. Concrete problems cannot, of course, be tackled by methodological considerations and postulates but only by patient investigation of the object. But on the other hand, fundamental attitudes determine what problems the investigator is able to see; they decide the framing of his questions, his experimental procedure, the choice of method, and finally, the type of explanation and theory that are given for the phenomena investigated. In fact, the dependence on prevailing attitudes of mind is the stronger the less it is felt. In this sense there is no doubt that the work achieved and the triumphs gained, as well as the shortcomings of classical biology, were determined by those leading principles which we have indicated. In order to realize this, it suffices to glance at any field of biology, and even of medicine and psychology as we shall see later. In a similar way, the organismic conception is a working attitude seeking to direct what problems shall be set and how they shall be solved. It makes it possible to see and to tackle basic problems of living phenomena and their possible explanations, problems that through the previous conceptions were either not seen at all or, if seen, considered to be mysteries inaccessible to scientific approach.

The aim is the statement of *exact laws*, which, according to the essential characteristics of living phenomena must, to a large extent, have the nature of system-laws. In this sense the organismic conception is a prerequisite for the transition of biology from the stage of natural history, i.e., description of forms and processes in the organisms, to an exact science. It seems to be the task which is set to our age, to accomplish in biology that "Copernican revolution" which, in the sciences concerned with inanimate nature, took place with the transition from the Aristotelian world-system to modern physics.

With this in mind, we will examine some basic biological problems and see how the organismic conception works. Thereafter we shall examine its epistemological consequences.

Edmund Sinnott

Edmund Ware Sinnott was an American botanist who was born in 1888. He was educated at Harvard University and received his Ph.D. in 1913. He taught at Harvard from 1908 to 1912 and subsequently at Connecticut Agricultural College, Barnard College and Columbia University. He was Sterling Professor of Botany at Yale University from 1940 to 1956, Director of the Sheffield Scientific School at Yale (1945-56), and Dean of the Yale Graduate School (1950-56). Sinnott was a member of the National Academy of Sciences, American Philosophical Society, and American Academy of Arts and Sciences. He was president of the American Association for the Advancement of Science in 1948. Among his works are: *Principles of Genetics* (with L. C. Dunn and Th. Dobzhansky), *Cell and Psyche, Two Roads to Truth, The Biology of the Spirit*, and *Matter, Mind and Man*. He died in 1968.

The following selection is taken from *Cell and Psyche*, (1950), pp. 15-42, and is reprinted here with the kind permission of the University of North Carolina Press, Chapel Hill.

CELL AND PSYCHE

Chapter I

Organization, the Distinctive Character of All Life

The universe is turning out to be a far more surprising place than our grandfathers ever dreamed. The more we learn of it, the wider grows the realm of the unknown. Science, like Hercules, is coping with a Hydra and finds that for every problem which is solved, two new ones rise at once to take its place. "An addition to knowledge," says Eddington, "is won at the expense of an addition to ignorance. It is hard to empty the well of Truth with a leaky bucket."[1]

In the last generation physics and chemistry and astronomy have completely rebuilt our old ideas about the world of nature. Ancient solidities and certainties have disappeared. Matter and energy and space and time have taken on quite other aspects and seem to be subject to analysis at last only by mathematical subtleties. During this same half century the sciences of life have also made great progress, especially in the application of physical and chemical knowledge and techniques to biological problems; but there has been no such revolution here as that which shook the physical sciences so profoundly. Biology is far more complex than they, and for it there has not yet come a formulation of the new and radical concepts which are necessary before life can truly be understood. Protoplasm still confronts us as the most formidable of enigmas.

But for all men life must nevertheless remain the ultimate problem. Around it, since we ourselves are living things, center those great questions which have always stirred mankind most deeply: on the lower level food, sex, race, and other problems of our animal nature; on the higher, those ultimate questions as to the place and significance of man in the universe, as to his personality and its destiny, his freedom, and the meaning for him of love and beauty, of virtue and aspiration, of what he calls his spirit and its communion with the universe outside of him. The structure of the atom, the size of space, and the theory of relativity interest a few, but rarely stir men deeply. No one goes to the barricades in defense of $E = mc^2$. But those more vital matters, which reach into our hearts as much as into our minds, have set wars ablaze and banners flying and poets singing and mystics praying since the dawn of history. These are all problems of *life*, and life is the ultimate mystery.

Any satisfying philosophy must deal with these questions, and to do so it must be rooted in the science of life itself, of life not only as we see it in man but as it is expressed in those far simpler organisms up and down the evolutionary scale. It is therefore biology in its widest sense, as the interpreter of life at every level, which will bring the richest offerings to philosophy. Tennyson's flower in the crannied wall, if we could really understand it, "root and all, and all in all," would indeed solve for us the final mysteries of God and man, for these are the mysteries of life itself.

What, then, can the biologist tell us about the curious phenomenon with which he deals? The nineteenth century produced the magnificent conception of life as dynamic, changing, ever moving forward; of the history of the world as the great stage on which the drama of organic evolution is being enacted. But it also established the equally important conception that life has its physical basis in that remarkable material system which is called *protoplasm*. Here in this aggregation of proteins—watery, formless, and flowing, deceptive in its visible simplicity but amazingly complex in its ultimate organization—are centered all the problems of living things. It is not greatly different chemically and physically in bacterium and orchid, in amoeba, arthropod, and man. Life is protoplasmic activity, and this is essentially the same from protozoan to primate. Man is not only cousin to all living things by blood-relationship, but is built of the very same stuff as they. It is not of dust or clay that we all are made, but of proteins and of nucleic acids.

The task of the biologist is therefore to understand this remarkable living material. From it are built the beautiful and intricate bodies of plants and animals; in it centers the control which regulates the activities of these exquisite mechanisms; and out of it come the alterations which make possible all evolutionary change. Early biologists believed that there must be some sort of soul or *anima* in every living thing, which governs it. A few, even in recent times, have been so much impressed with the complexity of protoplasmic activity, especially in its control of growth and development, that they adopt an essentially similar explanation and assume the existence of an entelechy or some other extraphysical agent which directs the activity of living stuff. Such a philosophy of vitalism, however, is now rarely asserted. Students of plant and animal physiology more commonly seek to explain in physical and chemical terms alone everything that goes on in protoplasm. The recent rapid growth of biochemistry has made it possible to analyze into relatively simple processes so many vital activities that this mechanistic view of biology has been greatly stimulated. It looks at life as simply a particularly complex series of physical and chemical reactions, no different fundamentally from those in any material system.

Protoplasm is a far more complicated affair, however, than biologists of a generation ago imagined it to be. An easy imitation, outside the organism, of some of the changes evident in living cells led them to the optimistic prediction that in a few years it would be possible even to synthesize protoplasm and produce a living thing. Such a triumph today seems farther away than ever. Physiologists have underestimated their protoplasmic opponent and have been obliged to withdraw, at least temporarily, from many advanced theoretical positions. Everything that we have learned about protoplasm in recent years testifies to a complexity in physical structure, chemical composition, and physiological activity within it far beyond that which its visible simplicity would lead us to expect; and when we realize that out of this remarkable stuff has come not only the protean plant and animal life of our globe but man himself, with the magnificent accomplishments and the sublime possibilities which are his, our respect for it should be profound. Protoplasm is a bridge anchored at one end in the simple stuff of chemistry and physics, but at the other reaching far across into the mysterious dominions of the human spirit.

A recognition of the magnitude of the problem which confronts them has far from discouraged biologists. For its solution they have enlisted the aid of their own clans—physiologists, morphologists, embryologists, geneticists, cytologists, microbiologists, and the rest—and have called in powerful allies from chemistry, physics, and mathematics. Their successes have been notable. The electron microscope has delved so deeply into protoplasmic structure that the genes themselves at last are visible. Some of the processes of metabolism, notably that of respiration, once thought to be fairly simple chemical exchanges, have been shown to involve many and complex steps and interactions and the mediation of a long series of enzymes. Growth and development in animals and plants are known to be affected by many chemical and physical factors—hormones, growth substances, organizers, bio-electric fields, light, temperature, and many others. Every living thing, even the humblest, is evidently a mechanism of the most remarkable and exquisite complexity.

What, we may ask, is the essential character of this mechanism, the quality that best distinguishes it? An obvious answer would be that it contains some substance or substances which make it what it is. This answer has often been given; and the increase in our knowledge of the chemical activities of living stuff and of the physiological importance of specific substances like the hormones has persuaded many biologists that the secret of life is indeed to be found in a persistent analysis of its biochemical behavior.

Others, however, who see the difficulty of this concept if it is carried very far have come to realize that it is not the *character* of the constituents of a living thing but the *relations* between them which are most significant. An organism is an *organized* system, each part or quality so related to all the rest that in its growth the individual marches on through a series of specific steps to a specific end or culmination, maintaining throughout its course a delicately balanced state of form and function which tends to restore itself if it is altered. This is the most important thing about it. E. B. Wilson in a famous passage said that "we cannot hope to comprehend the activities of the living cell by analysis merely of its chemical composition. . . . Modern investigation has, however, brought ever-increasing recognition of the fact that the cell is an *organic system*, and one in which we must recognize some kind of ordered structure or organization."[2] Woodger remarks that "biologists in their haste to become physicists, have been neglecting their business and trying to treat the organism not as an organism but as an aggregate. . . . If the concept of organization is of such importance as it appears to be it is something of a scandal that we have no adequate conception of it. The first duty of the biologist would seem to be to try and make clear this important concept. Some bio-chemists and physiologists . . . express themselves as though they really believed that if they concocted a mixture with the same chemical composition as what they call 'protoplasm' it would proceed to 'come to life.' This is the kind of nonsense which results from forgetting or being ignorant of organization."[3] Herbert Muller puts it well thus: "For the fundamental fact in biology, the necessary point of departure, is the organism. The cell is a chemical compound but more significantly a type of biological organization; the whole organism is not a mere aggregate but an architecture; the vital functions of growth, adaptation, reproduction—the final function of death—are not merely cellular but organic phenomena. Although parts and processes may be isolated for analytical purposes, they cannot be understood without reference to the dynamic, unified whole that is more than their sum. To say, for example, that a man is made up of certain elements is a satisfactory description only for those who intend to use him as a fertilizer."[4]

Through all the complexity which it is the task of the biologist to analyze thus runs one fundamental fact common to every living thing: protoplasm *builds organisms.* It does not grow into indeterminate, formless masses of living stuff. The growth and activity shown by plants and animals are not random processes but are so controlled that they form integrated, coordinated, organized systems. The word *organism* is one of the happiest in biology, for it emphasizes what is now generally regarded as the most characteristic trait of a living thing, its *organization*. Here is the ultimate battleground of biology, the citadel which must be stormed if the secrets of life are to be understood. All else are outworks, easily open to energetic attack. But this central stronghold, we must ruefully admit, has thus far almost entirely resisted our best efforts to break down its walls.

Organization is evident in diverse processes, at many levels, and in varying degrees of activity. It is especially conspicuous in the orderly growth which every organism undergoes and which produces the specific forms so characteristic of life. A coniferous tree, for example, such as a spruce or pine, though a loosely integrated plant individual in comparison to most animals, has a definite form, and its parts show a close coordination with each other. In each year's growth the central shoot is vertical and continues the axis of the trunk. The several side shoots, not as long, spread out almost horizontally. A definite pattern for the crown of the tree thus develops, the apical shoot growing faster than the branches, but the ratio between the two remaining essentially constant so that a regular conical shape is produced. If the young "leader" or terminal shoot is removed, one of the laterals swings up to take its place. This and other evidence indicates that the orientation and the relative growth of these side shoots are in some way under the control of the terminal bud. Other buds govern the growth of particular parts or branches. The angles which these make with the trunk, the ratio of height to diameter in the trunk itself, the proportion of above-ground parts to the root system, and other measurable relationships tend to be maintained. Thus the whole tree is an organized system in which the character and amount of growth in one part is related to that in all the others so that a precise form is reached. Some of the agents involved in this control, notably the plant hormones, are known; but how they are distributed so precisely in space and time that such a coordinated system is produced we do not understand. The tree itself is the expression of this organizing control.

A still more tightly organized system is evident in the developing animal embryo. The fertilized egg of a salamander is cleft into two cells by a vertical wall, then into four as one would quarter an apple, then horizontally into eight, and so on and on. If to our vision these changes are speeded up by time-lapse photography we can witness how the tiny group of cells, through continued cleavage, forms a partly hollow, spherical body; how the upper portion grows down over the yolk mass; how at one point the sphere is pushed in to make the primitive mouth; how above this the puckered neural folds mark out the position of the spinal axis; how they grow over to meet and form the neural tube; how at the sides the primitive gills appear; and how, step after step, the embryo moves swiftly on to form the young larva from which the mature salamander grows. Here is no random process but a steady march, each event in step with the rest as though to a definite and predetermined end. One gets an impression of some unseen craftsman who knows what he is about and who molds the mass of growing cells according to a precise plan. The young salamander seems to go through, before our eyes, an active progress toward a destination in a way which suggests its later movements of behavior, and not a merely passive unfolding. Here seems to be the expression in development of a constantly operating control which from the start and through all its precise steps from egg to adult maintains the embryo as an organized system.

This strict coordinated progression in organic growth is everywhere manifest, though often in less dramatic ways. The very fact that living things, in their bodies and in the organs which constitute them, everywhere show constant and specific *forms*, is proof of this control. Form is simply the external and visible expression of the organizing activity of protoplasm and is thus perhaps the most distinctive characteristic of living things. As Needham has well said, the central problem of biology is the form problem. In a gourd fruit, for example, growth in length and in width proceed at different rates so that form, as indicated by the ratio of one dimension to another, is continually changing. What remains constant is the ratio of the growth rates. During the development of the fruit any two rates keep evenly in step with each other so that it is possible to predict the actual dimensions and the changing dimensional ratios, and thus the organic pattern, at any stage of growth.

In the light of these facts it is impressive to look under the microscope at a thin slice of an early stage in such a developing fruit. Here one sees hundreds of tiny cells which by their constant division cause the organ to grow. The planes of these divisions—the angle of each new partition wall which cuts a cell into two—are in all directions. Chaos here seems to reign. This is no chaos, however, but a cosmos, with events marching to a precise destination, for the growth in the various dimensions of the organ which results from these divisions is beautifully coordinated. Some integrating control must here be operating. It is the nature of this control, of this fundamental organizing activity, which still eludes us and which constitutes the most formidable problem of biology.

One could multiply indefinitely examples of this sort, since all development normally shows such organized behavior; but among the lowliest of fungi there is an instance of this so remarkable that it illuminates the whole problem. In one group of slime molds (the *Ac-*

rasiaceae) the individuals are single cells, each a very tiny and quite independent bit of protoplasm resembling a minute amoeba. These feed on certain types of bacteria found in decaying vegetable matter and can readily be grown in the laboratory. They multiply by simple fission and in great numbers. When this has gone on for some time a curious change comes over the members of this individualistic society. They cease to feed, divide, and grow, but now begin to mobilize from all directions toward a number of centers, streaming in to each, as one observer describes it, like people running to a fire. Each center exerts its attractive influence over a certain limited region, and to it come some thousands of cells which form a small elongated mass a millimeter or two in length. These simple cells do not fuse, but each keeps its individuality and freedom of movement. The whole mass now begins to creep over the surface with a kind of undulating motion, almost like a chubby worm, until it comes to a situation relatively dry and exposed and thus favorable for spore formation, where it settles down and pulls itself together into a roundish body. Now begins a most curious bit of activity. Certain cells fasten themselves securely to the surface and there form collectively a firm disc. Others in the central axis of the mass become thick-walled and form the base of a vertical stalk. Still others, clambering upward over their comrades, dedicate themselves to the continued growth of the stalk. Up this stalk swarms the main mass of cells until they have risen several millimeters from the surface. These cells, a majority of the ones which formed the original aggregate, now mobilize themselves into a spherical mass terminating the tenuous stalk, which itself remains anchored to the surface by the basal disc. In this terminal mass every cell becomes converted into a rounded, thick-walled spore which, drying out and blown away by the wind, may start a new colony of separate amoeba-like cells. In other species the structure is even more complex, for the ascending mass of cells leaves behind it groups of individuals which in turn form rosettes of branches, each branch terminating in a spore mass. In this process of aggregation, a group of originally identical individuals is organized into a system wherein each has its particular function and undergoes a particular modification, some cells to form the disc, others the stalk, and others serving as reproductive bodies.

Such an aggregation of distinct cellular individuals into an organized system may also be observed in certain sponges. The living part of the body of such animals, consisting of at least four different kinds of cells, can be broken up artificially and even passed through muslin, but the thoroughly disorganized mass of cells, if they are not injured in the process, will regroup themselves in proper positions and produce a whole animal again. In some respects even more remarkable is a process which takes place among many insects, where the tissues of the caterpillar are broken down during the pupal or cocoon stage into what appears to be a disorganized mass of "mush." Out of this unpromising material the entirely different tissues and organs of the adult insect are gradually mobilized, a metamorphosis indeed, and one of the enigmas of biology.

Such organizing behavior is somewhat different from that in most plants and animals since here all growth (increase in material) is finished before differentiation and development begin, but we can hardly doubt that the process which integrates this group of individuals or a mass of homogeneous material and transforms it into an organized biological society is the same as that which operates in the more familiar cases of growth and development by cell multiplication. In both there is the same orderly progression, the same close coordination of one part with the rest, the same march to a final goal. In both, to use Driesch's famous dictum, "The fate of a cell is a function of its position." In both, there is the same evidence of unifying control. Surely if we could understand what makes the tiny cells of a slime mold run together and build such a beautifully fashioned cell-state, where each is modified in a particular way which serves the whole, we should know much about the ultimate secret of life.

The evidence of biological organization from these examples of normal growth and development is greatly extended through studies by which these processes are experimentally modified, especially by removing certain parts of the growing body. When this happens the organism shows a remarkable ability to regenerate its lost parts and restore a normal whole. Thus a "cutting," removed from a plant, under proper conditions will produce a new root system and finally an entire individual with the normal proportion of root to shoot. Internal plant structures may also be restored. If a conducting bundle in the growing stem or leaf is severed, the two ends may be connected by the development, behind the cut, of a new bundle through the conversion of ordinary storage cells into specialized vascular ones.

Some of the most remarkable examples of regeneration occur in animal embryology. Where the egg of a sea urchin or a frog, for example, at the beginning of development divides into two cells, these may be separated from each other, and each, instead of producing *half* an individual, now grows into a *whole* one. The fate of each cell is now quite different from what it would have been if it had remained part of a two-celled embryo. By the reorganization of its material each regenerates a single whole animal. Such behavior, of which countless similar examples might be cited, is so difficult to explain on chemical or physical grounds that Driesch, less tough-minded than most biologists, was driven to assume the operation here of an entelechy or director.

Regeneration is common everywhere in young, growing organisms. The leg of a tadpole, snipped off, may be restored, or the eye of a crustacean. Mature animals also may regenerate, as in the familiar case of the angleworm in which, when the body is cut in two, the head end will form a new tail. Regenerative ability is by no means universal, however, and is lost in most adult individuals or structures. In less highly organized systems, like most plants, it persists in certain more embryonic parts. Many cases are known where a single cell, from the surface layer of a leaf or elsewhere, may be induced to start independent development and to form an entire new individual. The general conclusion, with all its far-reaching implications, seems justified that every cell, fundamentally and under proper conditions, is *totipotent*, or capable of developing by regeneration into a whole organism.

In all these cases of regeneration the molding, coordinating, organizing activity of living stuff is emphasized. Here, as in normal cases, the ultimate result, the goal toward which development seems to move, tends to be a single complete organism, whatever may have been the vicissitudes of its developmental history. The organizing ability of protoplasm thus shown so dramatically in the processes of growth and development has long excited the interest of biologists. To answer the problems which it poses is the task of the science of morphogenesis, which endeavors to mobilize evidence and techniques for their solution from most of the other biological disciplines and from the physical sciences, as well.

This same organizing control is evident not only in development but in the protoplasmic activities by which the life of the individual is maintained. Around a living creature is its unorganized material environment, a random mixture of many things. Certain of these, its food, are continually being pulled into the organism, where at once they lose their random character and are built into the organized structure of a living system. Every plant and animal thus acts as an incorporating center which brings organic order out of environmental disorder.

Such a living organism, however, is extraordinarily unstable and sensitive to external conditions. It is an open system, and matter is continuously passing into it and out of it. It is the seat of innumerable chemical and physical changes incident to vital activity. And yet the very continuance of its life depends on the maintenance of relatively constant conditions within it—of water content, acidity, oxygen supply, a definite concentration of certain specific substances, and many more. This is not merely an equilibrium, a balance between forces. It is what the physiologists call a "steady state," and to maintain it the expenditure of energy is required. Life *is* the maintenance of such a constant set of conditions, and death is the inevitable result of their dislocation. In such a complex and open system, the first requisite is evidently a means whereby the many activities are so regulated that the necessary balance is constantly restored as external and internal changes upset it, and the inevitable tendency toward disorganization is continually resisted. Here again, as in the processes of development, each part of the system must be closely tied to all the rest so that changes in one activity or in one region may be compensated by those in another. It is therefore very hard for a physiologist to study any one activity by itself, a fact which makes the practice of this science peculiarly difficult and has led to many erroneous conclusions. The particular level of physiological balance may change as development progresses, or as the environment is altered, but for each state or condition there is set up in the organism a norm or standard toward the maintenance of which its activity is constantly regulated.

The most conspicuous and best known of these physiological regulations are those in the higher animals, particularly the mammals, which must maintain a very constant internal environment. The precisely controlled bodily temperature of man and the warm-blooded animals is a common example of this. Equally important, though less familiar, is the maintenance of uniform concentrations of sugar and oxygen in the blood, and similar constancies. The control of these physiological processes is well described by Cannon in a notable book.[5] He proposes for this state of balance the term *homeostasis*. The way in which this is maintained under changing conditions, and the ability of the body to regulate its vital processes so very delicately, is surely one of the most remarkable phenomena displayed by living things.

Such regulations are familiar in man and the higher animals, where the mechanisms involved are chiefly the nervous system and glands of internal secretion. Living cells which are far less specialized, however, are also capable of such self-regulation. Among plants, for example, the hydrogen-ion concentration (acidity) in the sap of cells of a given tissue is often very closely maintained despite external change. The concentration of various dissolved chemical substances in particular cells may also be kept very near to a given level under widely varying external concentrations. These are essentially the same sorts of regulations as in homeostasis but involve no nervous mechanism.

A remarkable fact about organic regulation, both developmental and physiological, is that, if the organism is prevented from reaching its norm or "goal" in the ordinary way, it is resourceful and will attain this by a different method. The end rather than the means seems to be the important thing. The significance of such facts for an interpretation of biological organization is obvious.

The maintenance of an organized self-regulating system seems to be a general attribute of protoplasm, but such manifestations of organization as have here been discussed are not by any means a necessary accompaniment of all life. The beautifully coordinated living system sometimes suffers a grievous loss of organization. Tumors, cancers, malformations, and innumerable abnormalities of growth in plants and animals are evidence that the organizing control is sometimes relaxed. Its most radical modification is shown by certain types of cells which may be cultured indefinitely in a nutrient solution and there multiply and grow into shapeless masses of tissue. Such cells remain alive and show certain physiological regulations, since a complete lack of organization would bring death; but they are unable to produce a formed organism where each cell has its particular structure and function, depending on its place in the whole living system.

There are evidently various *levels* of organization, some of which are subordinate to others in a kind of hierarchy. A cell is one such level, and the processes which go on within it maintain a certain independence; but cells are organized into tissues, tissues are grouped into organs, and organs into individual organisms. This organization may be very loose, as in certain lowly plants where most of the cells are alike and the individual can hardly be distinguished from a colony; the mass may be more closely tied together, as in a tree, where there is an indefinite number of leaves and branches but a general pattern for the whole; or it may be very tightly organized, as in most individual animals.

Organization, however different in degree, is primarily a matter of *relations*. Harrison well describes it thus: "Particulate units at any level are not wholly independent of one another. The relations of particles are part of the system and it is their behavior in relation to one another that constitutes 'organization.' . . . No particle or unit can be clearly understood or its behavior predicted unless its reactions with others are taken into consideration."[6]

An understanding of how this organization is set up and maintained is the biological problem to which every other is subordinate and contributory. Whatever repercussions it may have upon other fields of human inquiry, it is thus primarily a task for the student of biology in the broadest sense and must be undertaken on his terms. These terms may have to be enlarged, and we may need to learn the use of new methods of attack upon the problem, but it is life that we are seeking to understand, and life is the province of biology. As Needham warns us, "Organization is not something mystical and inaccessible to scientific attack. . . . It is for us to investigate the nature of this biological organization, not to abandon it to the metaphysicians because the rules of physics do not seem to apply to it."[7]

There have been many attempts to solve the problem of organization. For some biologists this presents no difficulty, and is simply the question of how such a regulatory mechanism has arisen in evolution. During its long course, only those variations which were useful in survival persisted, and through this age-long trial and error the nice adjustments of part and process gradually were developed, by chance favorable mutations, until the present beautifully coordinated organic systems were produced. Surely, these men contend, organization must be something which has thus evolved. That it is not intrinsic in protoplasm is proven by the fact that it is often lost in cases of abnormal growth.

This evolutionary explanation is an obvious one, but it has its difficulties. It can hardly make clear, for example, how the power of regeneration could have been acquired. There seems little likelihood that all the great variety of injuries and losses which a plant or animal can now repair (including those produced experimentally and which almost certainly would never be suffered in nature) have occurred in its ancestry so frequently that natural selection has had a chance to develop organisms able to repair them. To account for correlative changes such as would be required in the development of a regulatory mechanism has always been a major difficulty for the theory of selection.

Organizing *relations* are easy to observe and measure but are very difficult to explain physiologically. It is much easier to deal with *substances*, and in attempting to understand organization biologists have therefore thought more often in chemical than in physical terms. They have frequently postulated specific formative materials, hopefully expecting that these in some way would translate themselves into organizing relationships. Particularly significant among such are the various growth substances, regulators, and hormones which in recent years have been so intensively studied in plant and animal physiology. Among plants, for example, the effects of auxin have been found to be very numerous and important. It is concerned with cell enlargement, cambial activity, bud inhibition, root formation, leaf fall, and other activities, and thus markedly affects the development of the plant. But it is evident that auxin cannot do all these things by itself. It is the agent, the messenger, by which they are accomplished; but the beautifully coordinated results must come from the presence of just the right amount of auxin, at just the right place, and at just the right time. Something must control the auxin, must act as the headquarters from which the chemical messengers are dispatched. Here is the real problem. "When we discover," says J. S. Haldane, "the existence of an intraprotoplasmic enzyme or other substance on which life depends, we are at the same time faced with the question how this particular substance is present at the right time and place, and reacts to the right amount to fulfill its normal functions."[8] Moreover the secret of the action of such a substance lies not primarily in itself but in the specific organization of the cells upon which it acts. Auxin no more makes roots than a nickel makes a tune in a juke box. It simply sets in motion the activity of an organized system. Not the nickel or the auxin holds the secret, but the structure of the system itself.

The amphibian "organizer" postulated by Spemann is an example of the same difficulty. A bit of the roof of the primitive mouth of the young salamander embryo grafted almost anywhere on the body of another embryo will start a new embryonic axis and thus may make the animal a double one, like a Siamese twin. This bit of living tissue was thought to have in itself important organizing powers; but soon other agencies, simple chemical or physical factors, were found which had essentially the same effect, and Spemann himself finally admitted that his "organizer" was but a stimulus, an evocator, and that the real problem of organization lies in the responding system, in the living stuff itself, and not in the trigger which sets this off. Such chemical explanations of organization, despite the enthusiasm with which they have been sought, have not thrown much light upon the problem. Probably not many biologists today would agree with Julian Huxley's optimistic prediction in 1933 that we were then on the verge of reducing the organizing powers of a living thing to a chemical formula and storing it in a bottle. The beautiful structure of chemical molecules, especially in the proteins with their great size and complexity, has suggested that the form and organization of a living thing may in some way be determined by that of the specific proteins it contains. It is hard, however, to picture a mechanism which would bring this about. Baitsell and others have gone even further and suggested that the organism is itself a gigantic molecule and that the forces which integrate it are the same as those which hold together and organize atoms.

Biophysicists have also offered their explanations. Gurwitsch, impressed by the fact that the fruiting bodies of many fungi, constant and specific in their forms, are produced by a tangled mass of apparently similar fungus threads, believes that a formative "field" exists around the developing structure. Whence this arises and how it operates he is not clear. This general criticism can be made of most field theories proposed by other biologists. More concrete, however, is the suggestion of Burr and Northrop,[9] who believe that the secret of organization lies in the presence of a characteristic bioelectrical field in and about a living individual, which controls its development. They state the problem in terms of the physics of fields rather than of particles. This is a stimulating idea and well worth developing, but it is difficult to picture exactly how it operates in terms of what we now know about the activities of living things.

Physiological regulation is better understood than that in development and is known to be related to the activity of specific chemical substances. This regulation is extremely delicate, as any one administering insulin well knows, for there is always danger from too much or too little. The normal system, however, controls blood sugar automatically and with beautiful accuracy, an extraordinary accomplishment considering all the things that might go wrong and upset it. We are reminded of Henderson's remark that "sooner or later . . . we come upon the fact that a certain organ or group of cells accomplishes that which is requisite to the preservation of the equilibrium, varying the internal conditions according to the variation of the external conditions, in a manner which we can on no account at present explain."[10]

One of the most spectacular attempts to account for organic regulation has recently come from the engineers. Automatically controlled machines have long been familiar, but their complexity has been raised to an extraordinary degree in the production of the electronic calculator. This is a truly amazing device consisting of thousands of radio tubes connected in a complex fashion by which, almost instantly, huge sums can be manipulated and calculations made which would take a corps of computers years to perform. Such a calculator can store information for later use and thus possesses the rudiments of memory. The principle on which it is built may make possible, its inventors believe, the construction of a machine which will answer abstruse questions and may be said to display some degree of ability to reason. Properly constructed it might even play a moderately good game of chess! Dr. Wiener[11] has shown the marked similarities between the behavior of such a machine and that of the nervous system and believes that the key to a knowledge of the latter lies in the principles developed in these calculators and especially in the so-called "feed-back" mechanism. We must salute those who have built machines which have such fantastic possibilities for the service of man. One may question, however, whether these artifacts really give us more than an instructive analogy with protoplasmic regulation. After all, we are not made of tubes, wires, and gears, but of protein molecules. Our bodies are a triumph of chemical, not mechanical, engineering. The electronic calculator may grow into an accomplished robot, but one doubts if it can have an original idea or write a beautiful sonnet, as protoplasmic systems can.

We must frankly admit, I think, that, despite our ingenious experiments and speculations, no adequate explanation of biological organization is forthcoming. Despite all the advances in a knowledge of physiology and of the physical and chemical character of living stuff, such a solution seems to be almost as far away as ever. Biology has made enormous strides in the study of processes, of the successive series of chemical changes which go on in protoplasm; but these organizing relations which living things display present a much more formidable problem, and it may be that some new idea, some great generalization comparable to that of relativity for physics, will be necessary before we shall be able to understand the true nature of protoplasmic systems, so deceptively simple to outward view but the seat of that complex organized activity which is life.

The fact of organization has so impressed some biologists that they are even inclined to rank it as one of the basic facts in the universe. Thus L. J. Henderson, a biochemist who thought deeply in these matters, says, "I believe that organization has finally become a category which stands beside those of matter and energy."[12] Needham, in somewhat the same vein, writes: "Organization and Energy are the two fundamental problems which all science has to solve."[13] This is not far from the concept of complementarity proposed by Bohr. The important implications of these ideas are obvious.

Our problem, though first the task of the biologist, must evidently transcend his domain and enter that of philosophy. The list of philosophers who have undertaken to deal with it is considerable. Most notable among them, perhaps, is Whitehead, who based an important part of his system upon the fact of organization, not only in living things but throughout the universe. Biology for him is the study of the larger organisms and physics that of the smaller ones. The notion of particle he would replace by the notion of organism.

Whatever we may think of these deep matters, it is evident that organization as one sees it in living things is a very real fact, explain it how we will. In any problem dealing with life it must be taken into account. The hypothesis which I wish to propose here is that in the regulatory and organizing processes in protoplasm lies the foundation of what are called the psychological or mental activities in animals and especially in man. From a study of it some interpretations will suggest themselves which may help toward the solution of those great problems which were posed at the beginning of our discussion.

NOTES

[1]A. S. Eddington, *The Nature of the Physical World*, p. 229.

[2]E. B. Wilson, *The Cell in Development and Inheritance*, p. 760.

[3]J. H. Woodger, *Biological Principles*, pp. 281, 290.

[4]Herbert J. Muller, *Science and Criticism*, p. 107.

[5]W. B. Cannon, *The Wisdom of the Body*.

[6]R. G. Harrison, *Cellular Differentiation and Internal Environment*, Publication 14, American Association for the Advancement of Science, p. 77.

[7]Joseph Needham, *Order and Life*, pp. 7, 17.

[8]J. S. Haldane, *The Philosophic Basis of Biology*, p. 79.

[9]H. S. Burr and F. S. C. Northrop, "The Electro-dynamic Theory of Life," *Quarterly Review of Biology*, X (1935), 322-33.

[10]L. J. Henderson, *The Order of Nature*, p. 86.

[11]Norbert Wiener, *Cybernetics*.

[12]L. J. Henderson, *The Order of Nature*, p. 67.

[13]Joseph Needham, *Time: The Refreshing River*, p. 33.

E. S. Russell

Edward Stuart Russell was born in 1887. He was president of the Linnean Society of London and from 1921 to 1945 was Director of Fishery Investigations in the Ministry of Agriculture and Fisheries. Russell was an officer in the Conseil Permanent International pour l'Exploration de la Mer from 1930 to 1935. His best known book is *The Directiveness of Organic Activites*. Russell died in 1954.

The following selection is taken from *The British Journal for the Philosophy of Science*, Vol., I, No. 2, August 1950, pp. 108-116, and is reprinted here with the kind permission of Cambridge University Press.

THE "DRIVE" ELEMENT IN LIFE

I

If we look at living things quite simply and objectively we cannot but be struck by one feature of their activities, which seems to mark them off from anything inorganic. This is the active, persistent and regulatory nature of these activities. In this short article I shall try to illustrate and defend the thesis that there is common to all living things this basic element of directive striving, usually unconscious and blind, only rarely emerging into consciousness to become intelligently purposive. The master-end towards which the directive and persistent activities of the individual life converge is the completion of the life-cycle, including as a rule reproductive preparations for the repetition of the life-cycle in the ensuing generation. This thesis I stated on a former occasion in the following terms. Emphasising the essentially *active* nature of life, I wrote: "Structuro-functional wholeness or integrity, and specific structure, are actively built up and maintained in the course of development, chiefly by the morphogenetic and behavioural activity of cells or groups of cells. If this integrity is disturbed by injury or adverse environmental influences, it is, so far as possible, restored by appropriate physiological or morphogenetic activities on the part of the organism and its cells, so that the normal state is restored or a new adaptive norm of structure and function set up. The organism actively seeks out and selects the substances necessary for its metabolism, or draws them from its stores. It actively seeks in many cases its appropriate environment, and strives to maintain itself therein; it actively seeks in many cases a suitable ecological niche for its eggs and offspring. In all these ways, and in many others, the organism strives to persist in its own being, and to reach its normal completion or actualisation. This striving is not as a rule a conscious one, nor is there often any foresight of the end, but it exists all the same, as the very core of the organism's being."[1]

An array of facts was set out in the book containing the passage in justification of this thesis, and I shall not attempt to repeat or summarise them. Instead, I shall adduce some further instances in support, and then consider briefly what may be the relation between our own experience of striving and the unconscious drive-like, "triebartig," character of organic activities, both morphogenetic and behavioural.

2

One aspect of the active striving nature of life is familiar under the guise of the "struggle for existence." All good observers of animals and plants in their natural surroundings, from Charles Darwin downwards, have emphasised the strenuous character of the struggle for life and the struggle for reproduction. A classical example is the persistent growth towards the light of the trees and creepers in a tropical forest, vividly described by Hingston in the following passage: "Like the trees, the lianas are also struggling to get upward. Too weak of themselves, they need the shoulders of others. Ever on the look out for some object to grip, they curl or twine themselves around their victims, or shoot forth roots that cling into the bark, or hook themselves to objects by curved spines, or stretch out arm-like projections that grip the sides of the stems and trunks. In the end by some contrivance or another they get up and free themselves into the light. They are in a new world and a transformation follows. The dried-up twisted leafless cables shoot out branches in every direction, a tangled mat develops with profusion of foliage, smothering the tree up which it has scrambled and extending into the mats and branches of its neighbours, covering perhaps a dozen of them with its leafy pall, and at last bursting into flower, the final purpose of its long climb."[2] This example illustrates nicely the double purpose of the climber's struggle towards the light—to satisfy its own needs for growth and self-maintenance and to produce its seeds in suitable surroundings, in a word to complete or fulfil its normal life-cycle. It must be borne in mind that the struggle for existence and persistence is not primarily a struggle against other organisms, an internecine strife with neighbours, but essentially an individual struggle to obtain the means of life and to reproduce the race. There is inevitably competition, and the weak or unlucky go to the wall, but this is a consequence of the individual *conatus* or striving towards life and reproduction.

The active persistent nature of growth and development is particularly well marked in plants, and there are many familiar examples—the persistent and oft-repeated growth of the dandelion from a stubbed root, the way in which it forces itself up to the light through an inch or so of tarred foot-path, the shooting up of new branches from a truncated oak, the extraordinary way in which the roots of a birch or a pine burrow into and grip the most unpromising rock surfaces and afford precarious foothold for the tree. In his description of the Canadian scene in *The Transplanted*, Frederic Niven well illustrates this, when he writes, "Still among boulders, boulders as big as cottages, were a few trees here and there, trees that spread long roots out to great length, roots that gripped the rock claw-like, and took small hold on dust in their crevices."

In animals the persistent and striving character of vital activities is shown most clearly in their behaviour, as in the drive to satisfy hunger or to mate. The strength of the sex drive is well known to us from personal experience, and the most casual observation of our domestic animals reveals the active, strenuous and continuing nature of their sexual and reproductive behaviour. Think of a dog pursuing a bitch on heat, or a broody hen intent on satisfying her drive to incubate a clutch of eggs. All this is familiar and obvious—so obvious that we are apt to forget its significance.

But in animals directive effort is manifest not only in their behavioural actions but also in their morphogenetic and physiological activities, which at all stages of the life-cycle maintain and restore norms of function and structure, and replace what is missing, as in the healing of wounds and in regeneration—within the limits of the possible. Such activities do not proceed smoothly and automatically, or with a machine-like rigidity towards a fixed end. They are, especially in difficult conditions, flexible, persistent and variable. This characteristic of persistency with varied effort is most conspicuous in behavioural action, of which many examples could be adduced. I have tried to summarise these special characteristics of organic directive activity as follows[3]—When the goal is reached, action ceases, but if it is not reached, action usually persists and is often varied, alternative modes of action being employed if the normal one fails to attain the goal. Accordingly the goal may be reached in different ways, the end-state or terminus of action being more constant than the mode of reaching it.

As the goals to which organic activities are directive and persistent are normally related to the main biological ends of development, self-maintenance and reproduction, the direction of these activities is towards the completion or fulfilment of the life-cycle. Life-cycle completion is indeed the master law governing all the activities of the organism, to which other laws of smaller scope, such as the law of need-satisfaction, are subordinate.

In normal conditions the activities of the life-cycle proceed in a stereotyped fashion, repeating in detail the life-cycle of previous generations, though not without active effort, as in an animal's struggle to obtain food or the growth-effort of a plant to reach the light. But when conditions are difficult, and especially when there is operational interference by man, the stereotyped course of life-cycle activities is upset, and adaptive responses become necessary if norms are to be restored and the life-cycle completed. In many organisms, if not in all, such adaptive or directive reactions to unusual or even unprecedented situations are forthcoming, though in different degree in different organisms.

Even in the earliest stages of embryogeny, if differentiation has not proceeded too far, regulatory activities will lead to the production of a normal embryo from a half or a quarter of the segmenting egg. This in itself is perhaps not very remarkable, and may even be susceptible of explanation in dynamic terms, as Bertalanffy[4] maintains. But there are other phenomena of embryonic regulation which seem to prove the existence of a drive to restore normality from abnormal beginnings, even if the drive be not fully successful. A good example is that described by Holtfreter, who writes: "Perhaps one of the most impressive illustrations of the puzzling complexity and at the same time of the 'sensefulness' of the individuation process is provided by the following experiment. Several blastoporal lips from early gastrulae of *Amblystoma punctatum* were cut out and exposed for about 10 minutes to alkali, which caused them to disintegrate into a heap of single, amoeboid cells. The disorganisation was carried still further by stirring and intermingling the free cells by means of a glass needle. When the suspension fluid was subsequently neutralised, the cells reaggregated into one or several spherical bodies, the cellular arrangement of which was of course quite different from the original one in the gastrula. During the following days there occurred a certain amount of reorganisation, for instead of retaining their random distribution, the cells performed directed movements which led to their sorting out and regrouping into two germ layers; all mesodermal cells tended to disappear from surface positions and to slip into the interior of the bodies, while the endodermal cells, in competition with ectodermal cells, went to establish continuous, well oriented surface epithelia. The internal mass of mesoderm became further segregated into the distinctly separated tissues of notochord, somites and kidney. The latter formed long coils of nephric tubules provided with nephrostomes and surrounded by blood capillaries. The notochord cells, though not united into a single straight cord, were, on the other hand, not freely dispersed but appeared in the form of a continuous, slightly dendritic organ. Finally, the skeletal muscle cells were grouped into somites, the arrangement of which was, however, irregular."[5] In this remarkable case, through directive cell-migrations, selective cellular adhesions and mutual inductive action, there was formed from quite abnormal beginnings an approximation to a normal organisation. The complex activities concerned have all the appearance of co-operative striving to reach a normal end in conditions quite unprecedented in the history of the race. The experiment recalls the somewhat similar phenomenon of the reconstitution of sponges from dissociated cells.

In many organisms the loss of a part of the body or of an organ leads to its replacement by regeneration, restoring thus the structuro-functional integrity which is necessary for continued life and the completion of the life-cycle. The activities concerned may be very complex, involving not a simple replacement but a complete remodelling of the remaining parts in such a way as to re-constitute a new whole, as in "morphallaxis" in planarians and polychaets.

It is noteworthy that organs may be regenerated which are not likely to be lost or destroyed in the ordinary course of events, but have been removed experimentally by surgical operation. There is the classical case of the regeneration (from the upper edge of the iris) of the lens extirpated from the eye of certain species of newt, a case which led G. Wolff, who discovered it, to postulate the existence in living things of a primary, not evolved, power of "purposive" or directive response. It has even been shown that in certain mammals, after the complete removal by surgical operation of the ovaries, new ovaries can be formed from non-ovarian tissues, and that this is followed in some cases by normal pregnancy.[6] Such a response to the loss of the ovaries could never have occurred in the whole history of the race.

Adaptive response, as in regeneration, compensatory hypertrophy and functional adaptation generally, is a phenomenon widespread in the organic realm, coming into play when the normal course of life-cycle activities is hindered or thwarted by unfavourable circumstances or by the experimental interference of man. It is shown in very varied degree, but the power is never completely absent.

In the life-cycle of most organisms adaptive response is directive towards satisfying the needs and restoring the norms of the individual itself, including its needs for reproduction. In highly organised insect societies, however, adaptive response may serve the needs, not of the individual insect, but of the colony or "social organism" as a whole. The late W. M. Wheeler, one of the best philosophical biologists, in the interesting parallel he drew between the organism and the insect society, well illustrated this remarkable fact. "If the worker personnel be removed from a young ant colony," he wrote, "leaving only the fertile queen, we find that this insect, if provided with a sufficiently voluminous fat-body, will set to work and rear another brood, or in other words, regenerate the missing soma (of the colony). . . . On the other hand, if the queen alone be removed, one of the workers will often develop its ovaries and take on the egg-laying function of the queen. In ants such substitution queens, or gynaecoid workers, are not fertilised and are therefore unable to assume their mother's worker and queen-producing functions. The termites, however, show a remarkable provision for restituting both of the fertile parents of the colony from the so-called complemental males and females."[7] These complemental males and females are less developed forms of the "royal" caste, and take on the reproductive functions of the true kings and queens when these are removed from the colony. In this way the maintenance, development and reproduction of the termite state is ensured.

How the needs of the hive are supplied in normal conditions by the directive activities of the worker-bees has been fully elucidated by the researches of K. von Frisch and his collaborators, and they have shown also in what a remarkable manner the workers overcome the difficulties in providing for these needs which they encounter when their normal procedure is drastically upset by experimental interference. In the ordinary course of events the workers go through a regular cycle of activities, which is closely correlated with the state of development of their salivary and wax glands. In the first period of a worker-bee's active life, up to about the 10th day, its main occupation is the care of the larvae, which it nourishes with the secretion from its highly developed salivary glands, together with honey and pollen which it draws from the storage cells in the combs. During its second period, from the 10th to the 20th day, the salivary glands atrophy, and the wax glands on the underside of the abdomen develop greatly. At this stage in the cycle it produces wax and builds cells as required, also it cleans the hive, stores the food brought in by the foraging workers, makes its first orientation flights, and for the last two or three days acts as a watcher at the threshold of the hive. In the third and last period of its life it takes on the duty of foraging for honey and pollen, and during this stage its wax glands atrophy.

By an ingenious method Rösch contrived to separate the young workers from the old gang; all the bees were first driven into one half of the hive (A) and shut off from the other half (B); then the hive was turned round $180°$; the younger bees remained in half A, intent on their domestic duties, while the older bees flew off on foraging expeditions; when they returned they flew to their accustomed entrance which however led them into the B half of the hive only. Thus half A was inhabited by younger workers only, while half B was the home of the older bees. The half containing the young bees was accordingly without foragers, and food stores were soon used up. After a couple of days some of the bees lay starving on the bottom of the hive, and some of the larvae had been torn out of their cells and sucked to satisfy the food-need of the young bees. But on the third day a remarkable change in behaviour was observed. Bees of only one to two weeks in age went out foraging and came back laden with food, this in spite of the fact that their salivary glands were fully developed and their normal task was the feeding of the brood. The needs of the hive and not their stage of bodily organisation determined this adaptive modification of their normal behaviour; their salivary glands rapidly atrophied.

The older bees based on half B had been deprived of the younger stock that normally act as nurses and feeders of the larvae, but some of them that still had salivary glands in a functional state stepped into the breach and fed the young, retaining their glands in action long beyond the normal time. Here again the needs of the hive were attended to, rather than individual needs.

In another experiment by Rösch a hive was robbed of its building workers, and put in a situation where the construction of new comb became urgently necessary. The building of new cells was undertaken by bees that had passed the normal age for this task and whose wax glands were in course of disappearing. The remarkable fact came to light, through microscopical examination, that in these bees the fat-body had pressed itself up against the reduced wax glands and caused them to re-develop. K. von Frisch, from whose summary I take this account of Rösch's experiments, makes an interesting remark apropos of this last finding, to the effect that "Man ist versucht, zu sagen: Der Wille regiert den Körper." But he cautiously adds, "Doch wir wissen nichts vom Willen der Biene und lassen das Rätsel ungelöst."[8]

3

We have now surveyed a number of examples, drawn from quite different fields, which illustrate what I have called the active, persistent and regulatory character of organic activities, as shown particularly in adaptive response, whether behavioural or morphogenetic. Now we as conscious subjects have direct and immediate knowledge or experience of this element of drive or striving towards biological ends which ap-pears from objective evidence to be a characteristic common to all or many organic activities. We are part of Nature, and there must there-fore be some relation between our experienced striving and that observ-ed in other organisms. What then can be the relation between this experienced drive and the objective "drive-character" or organic ac-tivities in general? Must we not think of them as two aspects—internal and external—of the same basic reality? Must we not postulate a gener-al, supra-individual hormé or drive in life, after the fashion of Schopenhauer's "Wille?" To do so would be to go beyond the bounds of science into the realm of metaphysical speculation. It is obviously impossible in a short essay to follow up and discuss this hormic theory of life, and I shall content myself with a few observations.

It is clear that there is nothing to be gained by trying to interpret morphogenetic and physiological activities in psychological terms such as are applicable to our own activity as conscious subjects. It is possible, within limits, to interpret the behaviour of animals in such terms, as Bierens de Haan[9] in particular has shown, but this mode of interpretation becomes increasingly doubtful and uncertain when we try to apply it to animals very different from ourselves, and appears highly hypothetical if applied to morphogenetic activities, though this attempt has been made by Agar.[10] Yet it seems certain that there is some element or factor in organic activities generally which cannot be a property of a purely physical system, and our only clue to its nature is that it must in some sense be psychological.[11] We know that we ourselves are psychophysical unities, and certainly not purely physical systems, and we have every right to extend this conception to organisms in general.

We have a clue to the nature of directive activity in our own experience of striving. This is only a partial clue, and it is liable to be a misleading one, for our behavioural activity, which alone is directly known to us, is for the most part of a unique kind, not found elsewhere in the organic realm. It is often guided by conceptual thought; we pursue ends already explicitly present in consciousness, and our behaviour is to that extent intelligently purposive. But by no means all our behaviour-life reaches this purposive level; much of it is rooted in unconscious desires and impulses, closely related with our bodily and physiological state, as for instance in the hunger or the sex drive. Such instinctive and often unconscious drives we share with other animals, and these drives, we know, are not purely psychical but psycho-physical. We have therefore *some* direct insight into the nature of one form of directive activity, namely instinctive behaviour. Now there is, as I have pointed out in my *Directiveness of Organic Activities*, a close analogy between instinctive behaviour and directive activity in general. They have the same characteristics of persistency with varied effort towards reaching goals that are normally related to biological ends, without foreknowledge of those ends. The conclusion therefore seems reasonable that in directive activity generally, and not only in instinctive behaviour, there must be an element which we can conceive only as psychological, as a function of a psycho-physical organism.

As I have already said, this is not the occasion to discuss the very difficult question of the nature of organism and of directive activity. I have been concerned only to indicate a possible line of approach.

NOTES

[1] E. S. Russell, *The Directiveness of Organic Activities*, Cambridge, 1945, p. 190.
[2] R. W. G. Hingston, *A Naturalist in the Guiana Forest*, London, 1932, pp. 34-35.
[3] *The Directiveness of Organic Activities*, p. 110.
[4] L. von Bertalanffy, *Das biologische Weltbild*, Bern, 1949, I.
[5] J. Holtfreter, in *Growth in Relation to Differentiation and Morphogenesis, Society for Experimental Biology, Symposia*, 2, Cambridge, 1948, p. 21.
[6] Evidence summarised by F. Wood Jones, *Habit and Heritage*, London, 1943, pp. 33-35.
[7] W. M. Wheeler, *Essays on Philosophical Biology*, Cambridge, Mass., 1939, p. 19.
[8] K. von Frisch, *Zts. f. Tierpsychol.*, 1937, I, 12.
[9] J. A. Bierens de Haan, *Die tierischen Instinkte*, Leiden, 1940.
[10] W. E. Agar, *A Contribution to the Theory of the Living Organism*, Melbourne, 1943.
[11] See R. S. Lillie, *General Biology and Philosophy of Organism*, Chicago, 1945.

Reginald O. Kapp

Reginald Otto Kapp was an English electrical engineer who was born in 1885. From 1935 to 1950 he held the Pender chair of electrical engineering at University College, London. He had a consultant engineering practice and had been concerned with the technical aspects of large-scale electrical supply systems. It was from this experience that he entered academic life at the University of London. Kapp was Dean of the University Faculty of Engineering and a member of the Senate. His interests extended far beyond the limits of his profession as evidenced by his book, *Science and Materialism.* He died in 1966.

The following selection, a lecture given to postgraduate students at the London School of Economics and Political Science, October 22, 1953, is taken from *The British Journal for the Philosophy of Science,* Vol. V, No. 18, August 1954, pp. 91-103, and is reprinted here with the kind permission of Cambridge University Press.

LIVING AND LIFELESS MACHINES

Terms of Reference

When a theme has become the subject of prolonged discussion certain arguments from both sides tend to become rather stereotyped and taken for granted. It then becomes necessary for such arguments to be re-examined, both for their relevance and for the conclusions to which they lead. Even the constructions that have come to be put on the expressions employed in the discussion need, from time to time, to be carefully reviewed. The present paper is an attempt at such a re-examination of the traditional approach to the question: What are the implications of any similarities and differences that there may be between living and lifeless machines?

The Traditional Arguments

The controversy about this question is philosophically important because of its bearing on the question: Whether living matter is or is not controlled by something that is (*a*) distinct from it, and (*b*) non-material. Those who say that it is so controlled give such names as Life, Mind, Soul, Entelechy, Élan Vital, Diathete to the influence claimed to exercise control. In doing so they attribute the structure and performance of living things to the *aided* action of matter on matter, for they claim that this non-material influence aids the course of material events. By contrast, those who deny that living things are controlled by any sort of non-material influence attribute their structure and performance to the *unaided* action of matter on matter. Believers in aided action are sometimes called dualists, while believers in unaided action are called monists. But rather than become involved in a discussion about the justification for these names I shall avoid using them here.

Traditionally, believers in unaided action argue as follows: "A machine does not contain any substance that one could describe by such a term as Life or Entelechy. The living organism is in all significant features a machine. Therefore there is no reason to postulate any such substance for the living organism." Recently a number of mechanical devices have been specifically quoted in support of this school. Among them are digital and analogue computers, apparatus so designed that it maintains some quantity at a predetermined value, battery operated trucks encased in a good imitation of the shell of a tortoise, and other ingenious devices. Each is said to support the theory of unaided action in respect of one of the performances of a living organism, the inference being that a sufficient number of such devices, sufficiently varied, would between them cover the whole of life's activities.

Traditionally, believers in aided action reject the view that living organisms are essentially machines. They enumerate sundry accomplishments of which living organisms are and machines are not capable. So it has come to be accepted without question by both sides that to say that a living organism is a machine amounts to saying that it results from the unaided action of matter on matter, and that the proper way of refuting this theory is to prove that it is not a machine.

What Is a Machine?

This seems to me most odd and I hope that it will not be thought paradoxical but merely obvious when I say that protagonists on both sides ought to use exactly the opposite arguments. Believers in aided action ought to insist that living organisms are machines, and believers in unaided action ought to insist that they are not. The error is, I think, that no attempt is made on either side to think out what characterises a machine. For one cannot decide whether a living organism is a machine, nor what conclusions to draw if it is one, unless one has a criterion by which to distinguish between what is and what is not a machine. To say that a living organism is a machine means, if it means anything at all, that it differs in some significant way from a structure that is not a machine. What is this difference?

It would be correct enough, but quite inadequate to describe a machine merely as a structure in which energy is converted from one form to another. For that holds true of any arbitrary collection of bits and pieces, from a motor car down to a piece of iron that rusts, a rock face that crumbles, or the surface of the sea when the wind disturbs it. It is equally true of any *part* of a motor car, or of an assembly consisting of the wheels, together with a portion of the road, half the house that the car is passing and the boots of a pedestrian. I bring in this absurd illustration in order to make clear, firstly, that the conversion of energy must be in some specific form and, secondly, that there must be a criterion by which to define the boundaries of anything that one would call a machine.

When those who wish to prove the theory of unaided action use an illustration they never choose rusting iron, crumbling rock faces, or disturbed sheets of water; they never choose a random collection of bits and pieces. They always choose something that performs a wanted operation, something of which one can define the purpose, be it only to maintain a pointer in a predetermined position, something that an engineer too would call a machine, something teleologically determined. If they did not choose something like this the illustration would not be at all convincing.

The Input and Output Functions

A characteristic of all man-made machines is that they serve as instruments of control. They are manipulated at more or less frequent intervals by an operator, who uses handles, levers, pedals, push buttons, thumbscrews, and similar controlling devices; and their performance is not a random one but the one that the operator contemplates while he is exercising the control. This characteristic has caused engineers to speak of the input function and the output function of some of their machines, of servo-mechanisms for instance. They would be quite precise if they applied the same terms to every machine they worked with. Both an input and an output function can be distinguished for digital and analogue computers, servo-mechanisms, guided missiles set to home on a target, toy tortoises, in short, for every one of the devices quoted in support of the theory of unaided action. And the terms input and output function, be it well understood, do not refer to energy. The input energy and the output energy are distinct from the input and output functions. They have different origins, different destinations, and follow different paths.[1]

To say that the performance of a machine requires an input function is to say that specific things must be done to the machine for it to give its characteristic performance. And every man-made machine is equipped with special devices by means of which an operator can do those things to it. In a motor car they are called the controls. Corresponding devices are an essential constituent of analogue and digital computers, of any machine that an operator so adjusts that it will maintain a pointer in a predetermined position, of a "robot" aeroplane, of the mechanical tortoises said to be analogous to living organisms. They are what I have just spoken of as the controlling devices. They serve as receptors for the input function.

Those who say that the structure and performance of a living organism are controlled by a non-material operator distinct from its body say, in effect, th . : its characteristic output function occurs only when there is or has been a corresponding input function, just as it does for a man-made machine. As with this, they claim, something specific is only done *by* the organism when something specific has been done *to* it. And the living machine, one should expect them to insist, must be equipped with devices analogous to the controls of a motor car, with receptors for an input function, by means of which the assumed non-material operator provides the control or, as I prefer to put it, the diathesis. In other words, one should expect them to insist that a living organism is essentially a machine.

Yet it is their opponents who do this. These opponents, the believers in unaided action, deny that a living organism is controlled by any agent analogous to the operator of a man-made machine; they deny the need for any mechanisms in a living machine that would correspond to the controls of a motor car. And yet it is they, and not the believers in aided action, who quote in support of their theory the resemblance of living organisms to certain machines.

This is why, as I remarked earlier, the position of the contesting parties is inverted. Those who ought to argue that living organisms are machines do not and their opponents do.

Anyone who believes in *aided* action ought, if he thought the matter out, to argue thus: "I support the hypothesis that a living organism does not exhibit its characteristic output function when subjected only to random forces. I claim that it must be controlled by some influence that is distinct from itself. I claim that, in technical terms, there is an input function whenever the organism exhibits its characteristic behaviour. The same is true of a machine. So the lifelike performance of this mechanical tortoise tends to support my hypothesis in so far as similar output functions justify the assumption of the reality of an input function for both."

Anyone who believes in *unaided* action ought, on the other hand, to argue thus: "I support the hypothesis that a living organism does exhibit its characteristic output function when left entirely to the random influences of its environment. I do not believe that it is controlled at all in the sense in which a man-made machine is controlled. I do not accept the view that there is anything that an engineer would recognise as an input function. In this respect every living organism differs basically from any device that human ingenuity can produce. This mechanical tortoise, for instance, does not work if left entirely to the random forces of its environment. It only exhibits its characteristic output function when I manipulate it in certain ways, when I operate certain controls and make certain adjustments. It would be quite wrong to jump to the conclusion that the lifelike performance of this ingenious device proves the need for an input function for a live tortoise. The distinction between man-made and living machines is basic; it is that there is an input function for the former and none for the latter."

This, I repeat, is how both sides ought to argue. But instead believers in aided action deprecate the machine analogy. And believers in unaided action insist on it. They attempt to prove that the living organism does not need an operator distinct from itself by pointing to machines that do. From the resemblance of a living organism to a machine that has an input function they infer that the living organism has none. Surely it can only be in the Colleges of Unreason, described by Samuel Butler, that the traditional arguments can have originated. It is only there that one might expect so much attention to be given to the output function, which is barely relevant to the controversy, while the input function, which is very relevant, is being assiduously ignored.

The Hypothesis That Living Organisms Are Automatic Machines

A bad advocate may yet have a good case. So let us consider next what facts and reasoning would support belief in unaided action.

To justify such belief one would clearly have to prove that there was no input function to a living organism. And to say that there is no input function is to say that the living organism is 100 per cent automatic in the sense of being a self-constructing, self-operating, self-repairing, self-maintaining machine. This is what believers in unaided action must and do say even when they try to prove it by means of experiments and demonstrations with machines that are definitely not self-constructing, not self-operating, not self-repairing, not self-maintaining.

Some man-made machines are more automatic than others. And the degree to which a machine is automatic is a measure of the amount of control and adjustment that must be applied to it. In a very automatic machine the number of separate controlling devices is small and these only need to be operated at rare intervals. The driver of a very automatic engine can leave it alone for long intervals after he has started it up, and many parts of the engine may work satisfactorily for years before they require any attention whatever. The driver of a motor car equipped with automatic gear change need do less things with his hands and feet than if the gear is not automatic. A robot could be left without attention for long periods. The input functions might be reduced to the occasional adjustments of a few thumbscrews and some words spoken into a microphone. A self-winding watch keeps good time if adjusted only every few years.

A machine that was 100 per cent automatic would, by definition, not even require that much input function. Whether such a machine could be produced in a factory is not relevant. One relevant question is whether it is possible in theory, whether the concept of a machine that is 100 per cent automatic is consistent with physical science. This is an interesting question but I do not propose to discuss it now.

The question that I do propose to discuss is the more concrete one, whether observation of living organisms tends to support or to refute the hypothesis that they are 100 per cent automatic. For this is, be it remembered, a hypothesis and as such in need of support. Moreover, it

is fundamental in the dispute and has not been discussed in the literature. The present position is that both disputants can at present have
recourse only to faith: on the one side to faith in the existence of non-
material influences and on the other side to faith in the powers and
accomplishments of matter. If this is not always appreciated it is because
of a human weakness for describing one's own faith as facts and one's
opponent's as mere hypothesis.

On the question, automatic or not automatic, I find again that the
traditional view, blindly accepted by both sides, is the very opposite of
the view to which facts of observation point. It always seems to be taken
for granted that only those performances of living organisms that are
subject to conscious control can conceivably be non-automatic. All
other, and particularly purely vegetative processes, are usually declared
to be 100 per cent automatic both by believers in aided and in unaided
action. And yet no facts of observation support the hypothesis that living
organisms are 100 per cent automatic; on the contrary, evidence available to everyone, as I shall show in a moment, strongly supports the
hypothesis that living organisms are nearly, though not quite, 100
percent non-automatic.[2] Before we come to this evidence let us consider
the very minimum physical constituents of any machine, be it living or
lifeless.

The Essential Constituents of a Machine

There are, and always must be, three indispensable constituents,
namely: moving parts, fixed parts, and a source of energy.

That some things move in any machine is obvious. In moving they
transmit forces, they exert pushes and pulls. It is equally obvious that
there must be something to push and pull against. This is the frame. So
any machine, living or lifeless, must contain some parts that are fixed
relative to others that move.

Among the moving parts in a motor car engine are pistons, connecting
rods, shafts, wheels, valves, levers. These are guided and retained by
portions of the frame. This latter includes the bedplate, holding-down
bolts, bearings, cylinders, cylinder covers, casings, struts, tie-rods,
containers for fuel and lubricants. The frame of a motor car engine must
be adapted to its function, which means that it must be strong, rigid and
durable. So must each individual moving part. It is designed so that it

may not bend or buckle, break or tear under the imposed stresses. Surfaces subject to wear must be hard and smooth.

The properties required of the three main constituents are often mutually exclusive. Joints between moving parts must be loose; they are often hinged. Joints between parts of the frame must be rigid; they are bolted, riveted or welded. Moving parts must be light and most of them have a small inertia; a well-designed frame is comparatively heavy. So far as possible the moving parts and frame are made of substances that do not burn easily; but this is exactly what the fuel must do. It lacks the property of durability essential for the moving parts and the frame, and has the property of inflammability lacking in them.

The Basic Difference Between Living and Lifeless Machines

A living engine must, as I have said already, also consist of moving parts, frame and fuel. And I want to point out how basically the constituent parts of a living machine differ from the corresponding constituents of any man-made one. One might think at this stage that such a comparison is trivial and of no great philosophical significance, and that a true philosopher would ignore it in favour of other and bigger distinctions, such as capacity for thought, capacity for reproduction, a sense of values, higher things. But perhaps the notion will obtrude, as I proceed, that the real reason for reluctance to give attention to these problems of mechanics may be that they introduce some awkward questions.

In a man-made machine it is easy enough to distinguish on a casual inspection between the three essential constituents. One would not mistake the petrol for the bedplate nor a rivet for a cam. One should expect, therefore, when viewing a living machine to be able to tell quickly which parts were moving ones, which the frame and which the fuel. The properties of the materials and the shape of the items ought to provide an easy clue. In a man-made machine the materials used and their shapes have to be distinctive enough if the engine is to have a good thermal efficiency, to be light, to work without backlash, to last for a sufficient time without repair.

Let us list then first the moving parts of the human body. They include, of course, the legs, the arms, the hands, the fingers. The head is another moving part; it can nod and turn. The upper and lower portions of the trunk can each move relative to the other. In fact one cannot name one major division of the human body that is not a moving part.

A glance at the skeleton shows a design well adapted to this. Joints are hinged and not riveted, bolted or welded, as they would be if mobility were not essential. Similarly soft tissues are designed to yield. When one examines the constituent parts of any soft organ one finds that these, too, are moving parts. The tongue, the heart, the walls of blood vessels, individual muscle fibres, nearly all components, right down below microscopic dimensions, are well adapted to the function of mobility.

Where then is the frame? Is the living engine a curious kind of machine built of nothing but moving parts? Certainly not. If it were it could not work. It would collapse. By the principles of mechanics there must be a frame. Where is it?

There is, of course, no mystery. At any moment certain parts of the body cease to be moving parts and, for that moment, develop the function of the frame. They are altered to serve the function. They then take up the pushes and pulls exerted by the parts that are, at that moment, moving ones. Whenever this happens the shape of each component is temporarily changed. In other words, a significant difference between a living and a lifeless machine is that for the latter the frame is permanent, while for the former it is temporary. It is an *ad hoc* frame. While one takes one step the stresses within the body are constantly changing and those parts used at the moment to take up these stresses are changed in shape accordingly. Components that are tie-rods at one moment become struts at another and connecting rods at yet another. While one word is being pronounced the tongue and lips change their shape and function many times. It is the same with all internal parts down to each tiny cell. Every component of a living organism is a machine that works in a strange way indeed. There is continuous redesigning of all its minutest component parts.

The Principle of Continuous Redesigning

This redesigning explains the high weight efficiency of living substance. The theme is a big one and I regret that there is not time now to discuss it adequately. Let me be content merely to point out that a man-made walking machine with a permanent frame would have to be so designed that this frame would take up stresses when the heel was touching the ground as well as those others that occurred when the toe was doing so. But the living machine is such that while the heel is touching the ground there is nothing capable of taking up those stresses that will be imposed at the next moments as the toe comes down. The frame will be correctly redesigned at the right moment for this and will then no longer have anything that would resist the stresses imposed by the posture just abandoned. Similarly a motor car must be permanently fitted with a brake as well as an engine. But an animal has no brake while it is running. It is redesigned so as to acquire one when it is coming to a stop; and at this moment it ceases to have an engine of propulsion, an engine designed for the function of running. Any boxing machine would have to be equipped both with thrusting and parrying devices. A human boxer has each when needed and not at other times. An arm is a device for delivering a blow at one moment and a shield at the next.

Thus the conversion of the living body from one kind of machine to another is as radical as the conversion of swords into ploughshares. And it goes on so long as the organism is alive. In engineering such redesigning would call for a new set of blueprints, a new set of calculations many times a second. Can anyone seriously think that digital and analogue computers, toy tortoises, guided missiles come anywhere near to being a true analogy? For a true analogy it would not suffice that the toy tortoises were made of soft material that the operator could easily bend and twist into new shapes. Devices for doing the twisting and bending would have to be incorporated in those soft materials. In other words many subsidiary machines would have to be provided. Their function would be to control the shape of the tissues from moment to moment. And the subsidiary machines would, in turn, have to be made up of sub-subsidiary machines and so forth down to molecular dimensions. The operator of a toy tortoise that was even remotely like a live one would be kept busy to bring about all the changes in the shape of the component parts that occur in the live one.

I have noticed that a fact of common observation is often not properly appreciated until it is given a name. So let me find one for this weight-saving method by which living organisms do not embody any component parts that are not required for the function of the moment. I shall call it the Principle of Continuous Redesigning. It is one of the most basic of biological principles and deserves, indeed, much more discussion than there is time for on this occasion.

To the illustrations of this principle already given, further ones could be added without limit drawn both from large scale and microphysiology. But even if time permitted, additional illustrations would do little to make the Principle of Continuous Redesigning more self-evident. After all, anyone can easily find an unlimited number of further illustrations for himself.

The Principle of Continuous Repair and Maintenance

Let us now turn our attention to the fuel. In a motor car this is petrol and it is stored in a tank. In the human body it is chiefly glycogen and is stored in the muscles, having been converted from glucose in the liver. The chemical processes during muscle activity include the combination of muscle protein with sodium. This protein is therefore another part of the fuel. So both the glycogen and the protein serve the double function of being fuel and being constituents of those muscle fibres that are at one moment moving parts and at another components of the frame. The living body is analogous to a motor car in which the chassis, brakes, cylinders, pistons, connecting rods, valves and bearings all contained combustible material, some of which was burnt whenever the driver placed his foot on the accelerator. By engineering principles such an engine could not work. Depleted of their substance supports would weaken and collapse, tie-rods would break, struts buckle; valves and pistons would cease to be a tight fit; every working part would develop backlash. A motor car built to the principles adopted by life would fall to pieces before it had covered a mile. Indeed an engineer could only say of the living body: "This is not built to last. This is built to wear out."

Any man-made machine in which surfaces of the frame and moving parts are subject to much wear requires correspondingly frequent attention. Wear is corrected by replacing what has been worn away. So it is with the living body. If matter is worn away at a rate vastly exceeding that in any man-made machine, it is replaced at a corresponding rate. To the Principle of Continuous Redesigning one must add the Principle of Continuous Repair and Maintenance.

A Permeating Input Function

What is true of control in time is equally true of control in space. Large portions of the volume of any man-made machine never need to have anything done to them. This applies, for instance, to the inside of a shaft or a bedplate. But for a living machine one cannot say that the inside of any morsel of tissue is not subjected to controlled change. Even bones are the sites of much and varied activity. The tiniest membrane in a cell must conform to this rule. There is, in my terminology, a micro-diathesis in all living substance. In more colloquial terms, the living machine is analogous to a motor car for which the controls are not limited to a few items, such as the steering wheel, pedals, clutch handle and so forth, but occur throughout the structure. If one could make a toy tortoise that truly resembled a live one, the input function would not only have to be continuous; the controlling devices that enabled the operator to supply the input function would have to be legion. And they would have to be distributed throughout the tissues. The things done *to* such a device would have to be permeating as well as unremitting. An operator who had to cope with such a machine would surely not say that it was 100 per cent automatic! One would have to go to the Colleges of Unreason to meet that interpretation.

Summary and Conclusion

To summarise. The available facts point to a more basic resemblance between living and lifeless machines than has generally been recognised on either side of the dispute. Both kinds of machine are distinguishable from any structure that is not a machine in two basic features. Firstly, as frequently stressed by others, both exhibit characteristic output functions. Secondly, though less often pointed out in the past, the facts suggest that input functions are necessary for both. The most fundamental difference between the two types of machine is in the nature of the input function. For a man-made machine it is applied only occasionally and at a few places. For a living machine one must assume, at least until opposing facts have been found, that the input function is (a) continuous (b) permeating, and (c) operating on a scale too small for any material manipulator, indeed on a scale at which a Clerk Maxwell demon would work.

Thus certain hard facts tell against the hypothesis that living organisms are 100 per cent automatic. And these facts suggest that the control to which they are subjected is not applied only to the organism's *conscious* activities, but to *all* of them including even vegetative ones. The activity of what I prefer to call a diathete is of a nature that one cannot grasp with one's imagination. Is there any need that one should?

It may be argued that these hard facts will one day be interpreted in a way consistent with the hypothesis of unaided action. Perhaps. But if it is to be so the hard facts must be faced and an interpretation that will save the hypothesis deliberately sought. Until it has been found the hypothesis of unaided action cannot be maintained with the support of, but in spite of, observable facts. Its basis must continue to be pure faith.

Whether this condition will appear attractive or unattractive I do not know. But I think one major obstacle to its ready acceptance may be difficulty in finding a satisfactory answer to the question: Where is the operator that is assumed to be in control of this continuous and permeating input function? It is therefore worthwhile to point out that this question cannot have a meaning. For if the facts oblige one to assume the activity of a non-material influence in control of any living substance, the facts also force one, for reasons that it would take too long to explain here, to assume that such an influence must be without location.

NOTES

[1]As a term that would cover both input function and output function, that would cover indeed every process calling for control, guidance, selection, discrimination, timing I, have suggested elsewhere the word "diathesis" (see *Science* v *Materialism,* London, 1940, and *Mind, Life and Body,* London, 1951). This term makes it possible to define a machine as a device for converting diathesis from one form into another. But I do not need to use the term here. The more familiar terms "input function" and "output function" serve the present purpose equally well.

[2]As the hypothesis has recently been supported by some distinguished scientists that properly designed machines could both think and reproduce their kind it may be advisable for me to make it clear that I propose, on this occasion, to discuss only observable performances. And I do not propose to discuss either capacity for thought or capacity for reproduction.

H. A. C. Dobbs

H. A. C. Dobbs wrote on the philosophy of science with emphasis on the philosophy of physics. In particular he published several articles concerning relations between conceptions of time in physics and psychology. He wrote this article while associated with the Ministry of Internal Defense and Security at Kuala Lumpur, Malaya. Dobbs died in 1969.

The following selection is taken from *The British Journal for the Philosophy of Science,* Vol. VII, No. 30, August 1957, pp. 140-50, and is reprinted here with the kind permission of Cambridge University Press.

DIATHESIS, THE SELF-WINDING WATCH, AND PHOTOSYNTHESIS

1 *Diathesis and the Self-winding Watch*

In various recent publications Professor R. O. Kapp has put forward a new form of vitalism suggesting the need for an immaterial agency, called a "Diathete," to explain the characteristic behaviour of living things contrasted with that of "lifeless" machines, such as servo-mechanisms, however cunningly the latter may be contrived. This diathete is conceived to be a sort of entelechy, without location in space but capable of exerting a causal influence over physical structures, which are located in physical space. It follows of course that a diathete does not obey the laws of physics; since these laws apply only to processes and entities which are located in physical space.

Kapp was led to postulate the existence of this immaterial entelechy, the diathete, in connection with the peculiar behaviour of living structures, through observing that the most striking and characteristic feature of this behaviour is controlled and selective response on the part of an organism, to the random forces and influences exerted by its environment. In other words the peculiarity of a living thing is that it has the capacity to produce an orderly controlled output in response to a set of random stimuli which constitute the input. Kapp has put this point in the form of a question: "Can random forces be caused to produce a specific event at specified moments of time?" Kapp's answer is that no inorganic structure, such as a lifeless machine, can be devised which will respond to an input of random forces and disorderly influences with a controlled output, without the intervention and assistance of a non-physical entelechy, the "diathete."

The primary object of this paper is to refute Kapp's contention, that it is necessary to introduce a non-physical agency of this kind, in order to enable random forces to produce controlled movements and specific events at specified times. I shall do this by citing a particularly simple example of an inorganic material mechanism, which is completely subject to the laws of physics, but which none the less contrives to use random forces so as to produce controlled movements.

In considering Kapp's theory one is mainly concerned with *the process resulting from a diathete's operations*, which process he calls "*diathesis*," rather than with the nature of a diathete in itself. Diathesis he suggests: "As a term that would cover both input function and output function, that would cover indeed every process calling for control, guidance, selection, discrimination, timing."[1]

Kapp conceives diathesis as operating in all living organisms between the input and the output. A diathete operates causally; but the causal laws governing its operations are not the laws of physics. For example in the case of the human organism he says: "the particular link in the chain of causation called *Mind* does something other links do not. It introduces diathesis. It does so by selecting from the unco-ordinated mass of sense-date those that make up an appropriate stimulus for a given response; by exercising guidance, control by introducing specified order into the course of events."[2] In other words, a diathete operates in the input of human sense-perception as a detector which selects, out of the chaotic mass of all the stimuli impinging on the organism from the environment, an orderly pattern of information, thus giving rise to a perception of the external situation. But diathesis also operates in the output stage to ensure an orderly response to this perception, in the form of controlled movement adapted to the needs of the external situation.

On Kapp's theory similar considerations apply to all situations in which changes occur in living organisms under the guidance and control of diathesis. In all such changes you have an input selected by the operations of a diathete, which causes a change in the output of the system. The efficient cause of the change in the system is an exchange of energy between the diathetic structure and its environment. But the effect of this energy exchange is determined in part by diathesis. "One of the causes of a change of state in the controlled element is therefore for every physical change an exchange of energy with the environment. . . . But it is only one of the causes. The other is that the element receives a supply of diathesis in the form of control of the movement when the change occurs. . . . Without the supply of diathesis the movement would have been a random and not a specific one."

That is to say, whenever a change occurs in a system, as a result of perturbations due to random energy exchanges with the environment, we cannot expect to find anything but random output, *unless* there has been an intervention in the process by the immaterial agency which Kapp calls a diathete. "So we are led," according to Kapp, "to look for a mechanism that can only exist in a living substance, that has a very specific . . . action.[3] The words "that can only exist in a living substance" sufficiently indicate the vitalistic assumption underlying Kapp's theory of diathesis. The suggestion is that a "diathete," an immaterial agency, not subject to the laws of physics, is required to explain any reaction in which random disturbances produce non-random results. It is this contention which I shall now refute by discussing the case of the self-winding watch.

A self-winding watch is essentially a mechanism for converting some of the energy of random movements of a human arm into the controlled movements of the components which go to make up the works of the watch. The self-winding watch achieves this by means of a "rotor" mechanism which responds selectively to certain preferred components of the motions of the human arm to which it is strapped (the input) so as to store potential energy in a coiled spring—the mainspring of the watch. This potential energy is then used to move a system of geared wheels, which deliver the output of the machine, viz. the motions of the hour, minute, and second hands. In one sense, of course, the movements of a human arm are not random but controlled; in part by conscious volition, and in part by the various involuntary neural processes which go on in the motor areas of the brain. But this fact is irrelevant in the present context: since the factors which control the movement of a human arm have in general no relationship to, or correlation with, the positions of the hour, minute, and second hands of the watch; and it is the controlled movement of the latter which is the purpose of the watch. In relation to the movement of the hands of a watch the movements of the human arm to which the watch is strapped are random.

A watch comprises two mechanical systems: (i) a *power* system, consisting of the mainspring and the chain of wheels which it drives; (ii) a *control* system, consisting of the escapement which ensures that the energy derived from the power system, through the unwinding of the mainspring, is drawn off at a precisely controlled rate, at specified moments of time. This is achieved through the jewelled escapement lever, machined in the shape of a miniature anchor, the flukes of which engage the teeth of the so-called "escape" wheel—the escape wheel being the last of the train of wheels powered by the mainspring which move the hands of the watch. The escapement lever oscillates, under the influence of the hairspring, in such a way as to release one tooth of the escape wheel at a time, alternately stopping and releasing the escape wheel. This causes the train of wheels geared to the power system, the mainspring system, and the watch hands themselves, to move intermittently, in precisely controlled jumps, rather than continuously. Thus a self-winding watch is a complete answer to the question which Kapp asks (when he attempts to establish the peculiar nature of diathesis, and which he considers cannot be answered by any inorganic mechanical system):

"Is a system possible in which a continuous and random movement of particles causes a specific effect intermittently and only at specified moments of time?"[4]

Kapp might contend perhaps that the movements of the particles constituting a human arm which winds a self-winding watch, though random, were not "continuous." It is of course true that the arm of a human organism, conceived as a macroscopic object, does not execute continuous movements; and that there are periods of time during which it is at rest in relation to the inertial frame of reference provided by the Earth. During such periods the spring of a self-winding watch will not be wound and it will receive no potential energy. But there are other periods when such a human arm *is* in continuous motion with respect to the centre of gravity of the earth (the origin of the inertial frame of reference), and we may fairly assume that during such periods (which are the ones during which the mainspring of the self-winding watch is wound) the constituent particles of the human arm are also in continuous motion. If this be granted it is clear that the self-winding watch affords a simple example of an inorganic mechanism which has all the essential characteristics which Kapp ascribes to a diathete. Moreover the self-winding watch is a mechanism which is completely subject to the laws of physics. It therefore serves to refute Kapp's theory of diathesis.

However, although the example of the self-winding watch refutes Kapp's theory as it stands, it does serve to draw our attention to certain remarkable features, in certain types of mechanisms, which may provide a clue to a fully mechanistic treatment of living structures. To see this it is necessary to examine significant features of the self-winding watch which make it a "eudiathetous" mechanism in Kapp's sense; in other words, those features which enable the self-winding watch to cause random movements of a human arm to produce specific events at specified moments of time. By doing so we can exhibit peculiar features of eudiathetous mechanisms which are analogous to certain organic processes in living structures. In particular we shall find an analogy between some of these significant features in the self-winding watch and some processes in photosynthesis, which may point the way towards the closer assimilation of certain mechanical and organic processes, thus helping us towards the ultimate goal of exhibiting all living processes as developments of processes already found in non-living mechanisms.

The significant features are as follows:

A *Power System* comprising, *inter alia,* the following:

(i) *A Random Source of Energy.* In the self-winding watch this source is the motion of the human arm. But in other cudiathetous mechanisms any source will qualify provided its components are *random:* for example the "white noise" of solar radiation.

(ii) *A Selective Mechanism which picks out certain preferred components from the random source of energy.* In the self-winding watch this is the "rotor" mechanism. This rotor arranges that when the motion of the arm to which the watch is strapped has a component of momentum in the direction of the line joining the figure "12" to the figure "6" on the dial, above a certain threshold, this component turns a wheel which winds the spring. The rotor mechanism will not respond to any component in the direction "3" to "9" or "9" to "3."

(iii) *A means of storing the energy derived from the preferred components selected from the random source of energy.* This is necessary if the random forces, which are the source of energy, are to be conserved, so as to produce "a specific" event at specified moments of "time" (Kapp's criterion of diathesis). In the case of the self-winding watch the work done by the source, the random movements of the arm, is stored as potential energy in the mainspring of the watch.

(iv) *A "Cut-out" mechanism to prevent the overcharging, by the selected components of the random source of energy, above a certain critical level at which the stability of the system is endangered.* Such a cut-out mechanism is provided in the self-winding watch by a release mechanism, which operates when the tension in the spring goes beyond a certain critical value. It ensures that until the tension falls again below the critical level the rotor mechanism will rotate idly, doing no work upon the mainspring. Thus information as to the state of tension in the spring has to be fed back into the cut-out mechanism, and this kind of feed-back is a general feature of all "cut-out" devices.

(v) *A control system for the effective delivery of the stored energy to the right place at the right time.* In the self-winding watch the controlled release of the potential energy is arranged by the escapement system described above. The hairspring of the escapement system receives, and stores, the information provided by the movement of the jewelled escapement lever, and converts this information into instructions to the mainspring system through the medium of the escape wheel.

2 *The Self-winding Watch and Photosynthesis*

To conclude this paper I shall mention certain aspects of the process of photosynthesis which exemplify some of the characteristically eudiathetous processes of the self-winding watch which I have mentioned above. It has been usual to represent the total process of photosynthesis by the following simple formula which is probably true to a first approximation:

$$CO_2 \text{ (Carbon Dioxide)} + H_2O \text{ (Water)} + \text{Light}/(CH_2O) + O_2.$$

But analysis has shown that two distinct systems are involved, one of which is essentially a *power* system requiring an external source of energy (solar radiation); and the other a process of *controlled* chemical synthesis, whereby the chemical energy derived from the power system is used to reduce carbon dioxide to other substances which are synthesised for use by the plant. Thus, photosynthesis, like the self-winding watch, can be seen to involve the two aspects of a *Power* system and a *Control* system, involving

(a) the photolysis of water molecules into free molecular oxygen and hydrogen in a combined form—the so-called "light" reaction for which energy, derived from components in the visible portion of the spectrum of solar radiation is necessary;

(b) the conversion of carbon dioxide molecules to reduced carbon compounds, out of which other biological processes in the plant cell can synthesise fats and proteins. This is the so-called "dark" reaction: since it does not require visible radiant energy for its operation.

Experiment has shown that the "light" reaction of photosynthesis—the *power* system—is controlled by the "dark" reaction; in that when radiant energy in the form of light is given continuously the rate of photosynthesis is limited by the rate at which this second reaction proceeds. In this respect, therefore, the first phase of the reaction may be compared to the self-winding mainspring system of a watch whereas the second phase of the reaction may be regarded as analogous to the escapement system of a self-winding watch.

The *Power system* in photosynthesis arranges that certain components are selected from the random source of energy, the "white noise" of solar radiation, through the medium of chlorophyll, to supply the energy necessary for the photolysis of water into molecular oxygen and hydrogen in a combined form, and for the reduction of carbon dioxide to carbohydrate and other compounds.

The *Control system* in photosynthesis consists in the living plant cell of the organised system of enzymes which regulate the rate at which the processes go, and ensures that the unstable intermediate products of photosynthesis do not revert before they have played their parts. When sunlight falls upon the chlorophyll molecules contained in the plant's chloroplasts, these molecules absorb photons from both the violet and red ends of the visible spectrum, and reflect photons from the green-orange region—hence the visible colour of yellow-green which the chloroplasts have. When radiant energy is absorbed by a substance the energy may be released in various ways. The absorbed energy may merely increase the temperature of the substance by increasing the random movements of its molecules about their mean positions, thermal energy depending upon the movement of whole molecules.

(Radiant energy absorbed in this way has no significant effect in photosynthesis.) Or the energy absorbed may be used to effect internal changes within the molecules by causing transitions between energy levels. As far as photosynthesis is concerned the significant effect of the absorption of energy is to raise the chlorophyll molecules to "excited" energy levels: the molecules which absorb "violet" photons being raised to a higher energy level than the molecules which absorb "red" photons. Molecules which are lifted to an excited energy level by the absorption of radiant energy as a general rule slip back to their ground state with the release of the same amount of radiant energy as they have previously absorbed by the process known as "fluorescence." When, however, these molecules are subject to a steady bombardment of photons of radiant energy, i.e. when light is given continuously, they will be reexcited again as soon as they have released energy: so that they will be kept in a state of practically continuous excitement.

A peculiarity of the chlorophyll molecule is that, while it absorbs radiant energy strongly in the violet region of the visible spectrum, it releases energy by fluorescence only in the red region of the visible spectrum. This fact is remarkable because the absorption of a violet photon raises the chlorophyll molecule to a higher level than the red-photon absorption level. The explanation according to the current quantum theory, is that the energy difference is accounted for by a "radiationless" transition, within the molecule, in the course of which certain constituents of the molecule receive kinetic energy, which is "free energy" in the thermodynamic sense. Thus the chlorophyll molecule may be regarded as a chemical "machine" for the conversion of radiant energy into free mechanical energy. It is this conversion, of absorbed radiant energy to "Gibbs free" kinetic energy within the molecule, which provides the necessary power to do the work of breaking the valence bond between hydrogen and oxygen in the water molecule. Thus, unlike other processes in which energy absorbed from sunlight by a coloured compound is dissipated subsequently in increased thermal energy, or through fluorescence, or through the chemical degradation of the substance (as when sunlight causes the fading of coloured dyes), the absorption of the components of sunlight by chlorophyll provides a source of "free energy" in the thermodynamic sense; which free energy is used to reduce a chemically inert substance to highly reactive constituents.

The action of chlorophyll, in selecting certain preferred components from the random "white noise" of solar radiation, and in utilising these components to yield "free energy," is clearly analogous to the action of the selective "rotor" mechanism of the self-winding watch, that utilises selectively certain preferred components of the random motions of the human arm, to provide the free energy required to drive the watch. Moreover, when the photo-chemical process of splitting water into its constituents, hydrogen and oxygen, is slowed down, or prevented, there is an increase of fluorescence: the chlorophyll then gets rid of more of the energy absorbed from violet components (part of which is converted to useful kinetic energy by the "radiationless transition"), through increased fluorescence in the red emission band. This connection, between variation in the rate of photolysis of water and rate of fluorescence of the illuminated chlorophyll molecule, implies a rather sensitive feed-back or "cut-out" mechanism, although fluorescence never disposes of more than a very small percentage of the radiant energy absorbed by the chlorophyll. It is clear, however, that some mechanism, other than increased fluorescence, must operate to dispose of surplus free energy under conditions of high illumination and low photochemical activity, so as to protect the plant cells from photodynamical destruction. One suggestion is that the plant cell in such conditions used part of its surplus free energy to convert carotenoids to phytol chains, and to attach these phytol chains to magnesium porphyrin rings, thus synthesising chlorophyll molecules, in replacement of those chlorophyll molecules photodynamically destroyed when the chlorophyll machine is, so to speak, "running light" (i.e. when the production of free energy is in excess of the requirements of the water photolysis). On this hypothesis surplus free energy generated by the radiationless transitions of the chlorophyll molecules, activited by the "white noise" of solar radiation, would be accounted for (1) by photodynamical destruction of some chlorophyll molecules, and (2) by synthesis of new chlorophyll molecules from the carotenoids and porphyrins present in the plant cell, these two processes being in dynamical equilibrium with respect to the prevailing conditions of illumination and photolysis of water. Such dynamic equilibrium would constitute a "cut-out" mechanism sufficient to protect the stability of the plant cell. Of course the existence of this suggested cut-out mechanism is a speculation suggested by the self-winding watch analogy set out above. But the hypothesis, though speculative at present, is clearly susceptible of experimental verification or disproof. The acceptance of the hypothesis (in consequences of favourable experimental evidence) would entail some revision of current ideas, about the role of the chlorophyll molecule in photosynthesis, according to which the chlorophyll molecule is regarded as a *catalyst*. According to the hypothesis sketched above, on the basis of the analogy to the self-winding watch, the chlorophyll molecule functions as a chemical *machine* (rather than a catalyst) for the generation of free energy from sunlight; and, therefore, like other machines is liable to suffer damage (even destruction) if it "runs light."

We may sum up this description of the operations of the "power-system" of photosynthesis by saying that its result is the splitting of water, a substance of low potential chemical energy, into two highly reactive constituents, oxygen and hydrogen, the energy necessary having been derived from the selection of certain preferred components from the "white noise" of solar radiation. The hydrogen formed in this way is not liberated in free molecular form (as is the oxygen), but is combined with some substance which has not yet been identified chemically, so as to constitute a hydrogen donor of high potential chemical energy. The sensitive feed-back relation, between the rate of photolysis of water and rate of fluorescence in the chlorophyll, suggests that the unidentified hydrogen donor may be chlorophyll itself, or a chlorophyll complex. This at present unidentified hydrogen donor substance so to speak "traps," temporarily, the hydrogen liberated from the oxygen by the photolysis of water, and stores it in a readily available form for the reduction of carbon dioxide. It also appears to function as a "cut-out" mechanism with respect to the power-system in that, as we have noted above, the rate at which the photolysis of water proceeds, in relation to the quantity of radiant energy absorbed by the chloroplasts, is determined by the rate of the "dark reaction" by means of which carbon dioxide is reduced by acceptance of hydrogen from the hydrogen donor.

Thus the power-system in photosynthesis, provided by the action of the chlorophyll molecule, has the four features we have noted above in connection with the power-system of the self-winding watch. These features can be tabulated in the following way:

So much for the analogy between the *power-systems* of the self-winding watch and photosynthesis. As regards the *control-systems* there is little that can usefully be said at present (apart from what I have already said in connection with the "cut-out" mechanism of the power-system), because of our ignorance of those processes of organisation within living cells that control the balance and rate of enzyme activity. Moreover, up to the present no one has succeeded in achieving a *complete* photosynthetic process *in vitro* without a living plant. This

(1) *Random source of energy*	Random motion of the human arm	White noise of solar radiation
(2) *A selective operator to detect and convert certain components from the source of energy to Gibbs free energy*	The "rotor" of the self-winding watch	The absorption spectrum, and the "radiationless" transitions, of the chlorophyll molecule
(3) *Storage of the energy derived from the selected components of the random source of energy*	The mainspring system of the self-winding watch	The unidentified compound which traps the hydrogen, freed by the photolysis of water
(4) *Cut-out mechanism to avoid oversupplying the storage system*	The cut-out release which prevents overwinding of the mainspring of the watch	The reaction of the hydrogen-trap upon the unknown mechanism for disposing of excess excitation energy not required for the photolysis of water

suggests that there are elements in the process which depend upon the organisation of living matter.

It may then be asked: What is the value of this analogy between the self-winding watch and the phases of photosynthesis which have been identified in the laboratory? The answer is that it is of some heuristic value to point out analogies between the structure and functions of non-living mechanisms and living organisms wherever these analogies can be established. For it is only by the resolute extension of mechanistic thinking to its utmost limits that we can hope to avoid a premature and unnecessary form of vitalism of which I believe Kapp's theory of diathesis to be one example. (Professor Eccles' "interactionist" theory, expounded in his *Neurophysiology of Mind,* may be another example.)

NOTES

[1] R. O. Kapp, "Living and Lifeless Machines," this *Journal*, 1954, 5, 94 n.
[2] R. O. Kapp, *Mind, Life and Body*, London, 1951, p. 124.
[3] R. O. Kapp, *Mind, Life and Body*, London, 1951, p. 183.
[4] R. O. Kapp, *Mind, Life and Body*, London, 1951, p. 153.

Erwin Schrödinger

Erwin Schrödinger was an Austrian physicist who was born in 1887. He was a student at the University of Vienna from 1906 to 1910. He taught at the University of Stuttgart and was a professor of physics at the University of Zurich and later at the University of Berlin. From 1936 to 1938 Schrödinger taught at the University of Graz in Austria, from where he was forced to escape to Italy and ultimately to the United States. He enjoyed a brief stay at Princeton, but soon joined the Institute for Advanced Study in Dublin. He retired in 1955. The Nobel prize in physics was awarded Schrödinger in 1933. He was the author of: *Four Lectures on Wave Mechanics, Modern Atomic Theory, Statistical Thermodynamics,* and *What Is Life?* He died in Vienna in 1961.

The following selection is chapters six and seven from *What Is Life?* (1944), and is reprinted here with the kind permission of Cambridge University Press.

WHAT IS LIFE?

Chapter 6
Order, Disorder and Entropy

Nec corpus mentem ad cogitandum nec mens corpus ad motum, neque ad quietem nec ad aliquid (si quid est) aliud determinare potest.[1]
SPINOZA, Ethics, Pt III, Prop. 2

A Remarkable General Conclusion From the Model

Let me refer to the phrase . . . in which I tried to explain that the molecular picture of the gene made it at least conceivable that the miniature code should be in one-to-one correspondence with a highly complicated and specified plan of development and should somehow contain the means of putting it into operation. Very well then, but how does it do this? How are we going to turn "conceivability" into true understanding?

Delbrück's molecular model, in its complete generality, seems to contain no hint as to how the hereditary substance works. Indeed, I do not expect that any detailed information on this question is likely to come from physics in the near future. The advance is proceeding and will, I am sure, continue to do so, from biochemistry under the guidnace of physiology and genetics.

No detailed information about the functioning of the genetical mechanism can emerge from a description of its structure so general as has been given above. That is obvious. But, strangely enough, there is just one general conclusion to be obtained from it, and that, I confess, was my only motive for writing this book.

From Delbrück's general picture of the hereditary substance it emerges that living matter, while not eluding the "laws of physics" as established up to date, is likely to involve "other laws of physics" hitherto unknown, which, however, once they have been revealed, will form just as integral a part of this science as the former.

Order Based on Order

This is a rather subtle line of thought, open to misconception in more than one respect. All the remaining pages are concerned with making it clear. A preliminary insight, rough but not altogether erroneous, may be found in the following considerations:

It has been explained in chapter 1 that the laws of physics, as we know them, are statistical laws.[2] They have a lot to do with the natural tendency of things to go over into disorder.

But, to reconcile the high durability of the hereditary substance with its minute size, we had to evade the tendency to disorder by "inventing the molecule," in fact, an unusually large molecule which has to be a masterpiece of highly differentiated order, safeguarded by the conjuring rod of quantum theory. The laws of chance are not invalidated by this "invention," but their outcome is modified. The physicist is familiar with the fact that the classical laws of physics are modified by quantum theory, especially at low temperature. There are many instances of this.

Life seems to be one of them, a particularly striking one. Life seems to be orderly and lawful behaviour of matter, not based exclusively on its tendency to go over from order to disorder, but based partly on existing order that is kept up.

To the physicist—but only to him—I could hope to make my view clearer by saying: The living organism seems to be a macroscopic system which in part of its behaviour approaches to that purely mechanical (as contrasted with thermodynamical) conduct to which all systems tend, as the temperature approaches the absolute zero and the molecular disorder is removed.

The non-physicist finds it hard to believe that really the ordinary laws of physics, which he regards as the prototype of inviolable precision, should be based on the statistical tendency of matter to go over into disorder. I have given examples in chapter 1. The general principle involved is the famous Second Law of Thermodynamics (entropy principle) and its equally famous statistical foundation. . . . I will try to sketch the bearing of the entropy principle on the large-scale behaviour of a living organism—forgetting at the moment all that is known about chromosomes, inheritance, and so on.

Living Matter Evades the Decay to Equilibrium

What is the characteristic feature of life? When is a piece of matter said to be alive? When it goes on "doing something," moving, exchanging material with its environment, and so forth, and that for a much longer period than we would expect an inanimate piece of matter to "keep going" under similar circumstances. When a system that is not alive is isolated or placed in a uniform environment, all motion usually comes to a standstill very soon as a result of various kinds of friction; differences of electric or chemical potential are equalized, substances which tend to form a chemical compound do so, temperature becomes uniform by heat conduction. After that the whole system fades away into a dead, inert lump of matter. A permanent state is reached, in which no observable events occur. The physicist calls this the state of thermodynamical equilibrium, or of "maximum entropy."

Practically, a state of this kind is usually reached very rapidly. Theoretically, it is very often not yet an absolute equilibrium, not yet the true maximum of entropy. But then the final approach to equilibrium is very slow. It could take anything between hours, years, centuries . . . To give an example—one in which the approach is still fairly rapid: if a glass filled with pure water and a second one filled with sugared water are placed together in a hermetically closed case at constant temperature, it appears at first that nothing happens, and the impression of complete equilibrium is created. But after a day or so it is noticed that the pure water, owing to its higher vapour pressure, slowly evaporates and condenses on the solution. The latter overflows. Only after the pure water has totally evaporated has the sugar reached its aim of being equally distributed among all the liquid water available.

These ultimate slow approaches to equilibrium could never be mistaken for life, and we may disregard them here. I have referred to them in order to clear myself of a charge of inaccuracy.

It Feeds on "Negative Entropy"

It is by avoiding the rapid decay into the inert state of "equilibrium" that an organism appears so enigmatic; so much so, that from the earliest times of human thought some special non-physical or supernatural force (*vis viva*, entelechy) was claimed to be operative in the organism, and in some quarters is still claimed.

How does the living organism avoid decay? The obvious answer is: By eating, drinking, breathing and (in the case of plants) assimilating. The technical term is *metabolism*. The Greek word ($\mu\epsilon\tau\alpha\beta\acute{a}\lambda\lambda\epsilon\iota\nu$) means change or exchange. Exchange of what? Originally the underlying idea is, no doubt, exchange of material. (E.g. the German for metabolism is *Stoffwechsel*.) That the exchange of material should be the essential thing is absurd. Any atom of nitrogen, oxygen, sulphur, etc., is as good as any other of its kind; what could be gained by exchanging them? For a while in the past our curiosity was silenced by being told that we feed upon energy. In some very advanced country (I don't remember whether it was Germany or the U.S.A. or both) you could find menu cards in restaurants indicating, in addition to the price, the energy content of every dish. Needless to say, taken literally, this is just as absurd. For an adult organism the energy content is as stationary as the material content. Since, surely, any calorie is worth as much as any other calorie, one cannot see how a mere exchange could help.

What then is that precious something contained in our food which keeps us from death? That is easily answered. Every process, event, happening—call it what you will; in a word, everything that is going on in Nature means an increase of the entropy of the part of the world where it is going on. Thus a living organism continually increases its entropy—or, as you may say, produces positive entropy—and thus tends to approach the dangerous state of maximum entropy, which is death. It can only keep aloof from it, i.e. alive, by continually drawing from its environment negative entropy—which is something very positive as we shall immediately see. What an organism feeds upon is negative entropy. Or, to put it less paradoxically, the essential thing in metabolism is that the organism succeeds in freeing itself from all the entropy it cannot help producing while alive.

What is Entropy?

What is entropy? Let me first emphasize that it is not a hazy concept or idea, but a measurable physical quantity just like the length of a rod, the temperature at any point of a body, the heat of fusion of a given crystal or the specific heat of any given substance. At the absolute zero point of temperature (roughly $-273°C$) the entropy of any substance is zero. When you bring the substance into any other state by slow, reversible little steps (even if thereby the substance changes its physical or chemical nature or splits up into two or more parts of different physical or chemical nature) the entropy increases by an amount which is computed by dividing every little portion of heat you had to supply in that procedure by the absolute temperature at which it was supplied—and by summing up all these small contributions. To give an example, when you melt a solid, its entropy increases by the amount of the heat of fusion divided by the temperature at the melting-point. You see from this, that the unit in which entropy is measured is cal./°C (just as the calorie is the unit of heat or the centimetre the unit of length).

The Statistical Meaning of Entropy

I have mentioned this technical definition simply in order to remove entropy from the atmosphere of hazy mystery that frequently veils it. Much more important for us here is the bearing on the statistical concept of order and disorder, a connection that was revealed by the investigations of Boltzmann and Gibbs in statistical physics. This too is an exact quantitative connection, and is expressed by

$$\text{entropy} = k \log D,$$

where k is the so-called Boltzmann constant (= $3 \cdot 2983 \cdot 10^{-24}$ cal./$^{\circ}$C), and D a quantitative measure of the atomistic disorder of the body in question. To give an exact explanation of this quantity D in brief non-technical terms is well-nigh impossible. The disorder it indicates is partly that of heat motion, partly that which consists in different kinds of atoms or molecules being mixed at random, instead of being neatly separated, e.g. the sugar and water molecules in the example quoted above. Boltzmann's equation is well illustrated by that example. The gradual "spreading out" of the sugar over all the water available increases the disorder D, and hence (since the logarithm of D increases with D) the entropy. It is also pretty clear that any supply of heat increases the turmoil of heat motion, that is to say, increases D and thus increases the entropy; it is particularly clear that this should be so when you melt a crystal, since you thereby destroy the neat and permanent arrangement of the atoms or molecules and turn the crystal lattice into a continually changing random distribution.

An isolated system or a system in a uniform environment (which for the present consideration we do best to include as a part of the system we contemplate) increases its entropy and more or less rapidly approaches the inert state of maximum entropy. We now recognize this fundamental law of physics to be just the natural tendency of things to approach the chaotic state (the same tendency that the books of a library or the piles of papers and manuscripts on a writing desk display) unless we obviate it. (The analogue of irregular heat motion, in this case, is our handling those objects now and again without troubling to put them back in their proper places.)

Organization Maintained by Extracting "Order" from the Environment

How would we express in terms of the statistical theory the marvellous faculty of a living organism, by which it delays the decay into thermodynamical equilibrium (death)? We said before: "It feeds upon negative entropy," attracting, as it were, a stream of negative entropy upon itself, to compensate the entropy increase it produces by living and thus to maintain itself on a stationary and fairly low entropy level.

If D is a measure of disorder, its reciprocal, $1/D$, can be regarded as a direct measure of order. Since the logarithm of $1/D$ is just minus the logarithm of D, we can write Boltzmann's equation thus:

$$- \text{(entropy)} = k \log (1/D).$$

Hence the awkward expression "negative entropy" can be replaced by a better one: entropy, taken with the negative sign, is itself a measure of order. Thus the device by which an organism maintains itself stationary at a fairly high level of orderliness (= fairly low level of entropy) really consists in continually sucking orderliness from its environment. This conclusion is less paradoxical than it appears at first sight. Rather could it be blamed for triviality. Indeed, in the case of higher animals we know the kind of orderliness they feed upon well enough, viz. the extremely well-ordered state of matter in more or less complicated organic compounds, which serve them as foodstuffs. After utilizing it they return it in a very much degraded form—not entirely degraded, however, for plants can still make use of it. (These, of course, have their most powerful supply of "negative entropy" in the sunlight.)

Note to Chapter 6

The remarks on *negative entropy* have met with doubt and opposition from physicist colleagues. Let me say first, that if I had been catering for them alone I should have let the discussion turn on *free energy* instead. It is the more familiar notion in this context. But this highly technical term seemed linguistically too near to *energy* for making the average reader alive to the contrast between the two things. He is likely to take *free* as more or less an *epitheton ornans* without much relevance, while actually the concept is a rather intricate one, whose relation to Boltzmann's order-disorder principle is less easy to trace than for entropy and "entropy taken with a negative sign," which by the way is not my invention. It happens to be precisely the thing on which Boltzmann's original argument turned.

Chapter 7
Is Life Based on the Laws of Physics?

Si un hombre nunca se contradice, será porque nunca dice nada.[3]
MIGUEL DE UNAMUNO (quoted from conversation)

New Laws to be Expected in the Organism

What I wish to make clear in this last chapter is, in short, that from all we have learnt about the structure of living matter, we must be prepared to find it working in a manner that cannot be reduced to the ordinary laws of physics. And that not on the ground that there is any "new force" or what not, directing the behaviour of the single atoms within a living organism, but because the construction is different from anything we have yet tested in the physical laboratory. To put it crudely, an engineer, familiar with heat engines only, will, after inspecting the construction of an electric motor, be prepared to find it working along principles which he does not yet understand. He finds the copper familiar to him in kettles used here in the form of long, long wires wound in coils; the iron familiar to him in levers and bars and steam cylinders is here filling the interior of those coils of copper wire. He will be convinced that it is the same copper and the same iron, subject to the same laws of Nature, and he is right in that. The difference in construction is enough to prepare him for an entirely different way of functioning. He will not suspect that an electric motor is driven by a ghost because it is set spinning by the turn of a switch, without boiler and steam.

But F. Simon has very pertinently pointed out to me that my simple thermodynamical considerations cannot account for our having to feed on matter "in the extremely well ordered state of more or less complicated organic compounds" rather than on charcoal or diamond pulp. He is right. But to the lay reader I must explain that a piece of un-burnt coal or diamond, together with the amount of oxygen needed for its combustion, is also in an extremely well ordered state, as the physicist understands it. Witness to this: if you allow the reaction, the burning of coal, to take place, a great amount of heat is produced. By giving it off to the surroundings, the system disposes of the very considerable entropy increase entailed by the reaction, and reaches a state in which it has, in point of fact, roughly the same entropy as before.

Yet we could not feed on the carbon dioxide that results from the reaction. And so Simon is quite right in pointing out to me, as he did, that actually the energy content of our food *does* matter; so my mocking at the menu cards that indicate it was out of place. Energy is needed to replace not only the mechanical energy of our bodily exertions, but also the heat we continually give off to the environment. And that we give off heat is not accidental, but essential. For this is precisely the manner in which we dispose of the surplus entropy we continually produce in our physical life process.

This seems to suggest that the higher temperature of the warm-blooded animal includes the advantage of enabling it to get rid of its entropy at a quicker rate, so that it can afford a more intense life process.

I am not sure how much truth there is in this argument (for which I am responsible, not Simon). One may hold against it, that on the other hand many warm-blooders are *protected* against the rapid loss of heat by coats of fur or feathers. So the parallelism between body temperature and "intensity of life," which I believe to exist, may have to be accounted for more directly by van 't Hoff's law, mentioned [in chapter 5]: the higher temperature itself speeds up the chemical reactions involved in living. (That it actually does, has been confirmed experimentally in species which take the temperature of the surrounding.)

Reviewing the Biological Situation

The unfolding of events in the life cycle of an organism exhibits an admirable regularity and orderliness, unrivalled by anything we meet with in inanimate matter. We find it controlled by a supremely well-ordered group of atoms, which represent only a very small fraction of the sum total in every cell. Moreover, from the view we have formed of the mechanism of mutation we conclude that the dislocation of just a few atoms within the group of "governing atoms" of the germ cell suffices to bring about a well-defined change in the large-scale hereditary characteristics of the organism.

These facts are easily the most interesting that science has revealed in our day. We may be inclined to find them, after all, not wholly unacceptable. An organism's astonishing gift of concentrating a "stream of order" on itself and thus escaping the decay into atomic chaos—of "drinking orderliness" from a suitable environment—seems to be connected with the presence of the "aperiodic solids," the chromosome molecules, which doubtless represent the highest degree of well-ordered atomic association we know of—much higher than the ordinary periodic crystal—in virtue of the individual role every atom and every radical is playing here.

To put it briefly, we witness the event that existing order displays the power of maintaining itself and of producing orderly events. That sounds plausible enough, though in finding it plausible we, no doubt, draw on experience concerning social organization and other events which involve the activity of organisms. And so it might seem that something like a vicious circle is implied.

Summarizing the Physical Situation

However that may be, the point to emphasize again and again is that to the physicist the state of affairs is not only not plausible but most exciting, because it is unprecedented. Contrary to the common belief, the regular course of events, governed by the laws of physics, is never the consequence of one well-ordered configuration of atoms—not unless that configuration of atoms repeats itself a great number of times, either as in the periodic crystal or as in a liquid or in a gas composed of a great number of identical molecules.

Even when the chemist handles a very complicated molecule *in vitro* he is always faced with an enormous number of like molecules. To them his laws apply. He might tell you, for example, that one minute after he has started some particular reaction half of the molecules will have reacted, and after a second minute three-quarters of them will have done so. But whether any particular molecule, supposing you could follow its course, will be among those which have reacted or among those which are still untouched, he could not predict. That is a matter of pure chance.

This is not a purely theoretical conjecture. It is not that we can never observe the fate of a single small group of atoms or even of a single atom. We can, occasionally. But whenever we do, we find complete irregularity, co-operating to produce regularity only on the average. . . . The Brownian movement of a small particle suspended in a liquid is completely irregular. But if there are many similar particles, they will by their irregular movement give rise to the regular phenomenon of diffusion.

The disintegration of a single radioactive atom is observable (it emits a projectile which causes a visible scintillation on a fluorescent screen). But if you are given a single radioactive atom, its probable lifetime is much less certain than that of a healthy sparrow. Indeed, nothing more can be said about it than this: as long as it lives (and that may be for thousands of years) the chance of its blowing up within the next second, whether large or small, remains the same. This patent lack of individual determination nevertheless results in the exact exponential law of decay of a large number of radioactive atoms of the same kind.

The Striking Contrast

In biology we are faced with an entirely different situation. A single group of atoms existing only in one copy produces orderly events, marvellously tuned in with each other and with the environment according to most subtle laws. I said, existing only in one copy, for after all we have the example of the egg and of the unicellular organism. In the following stages of a higher organism the copies are multipled, that is true. But to what extent? Something like 10^{14} in a grown mammal, I understand. What is that! Only a millionth of the number of molecules in one cubic inch of air. Though comparatively bulky, by coalescing they would form but a tiny drop of liquid. And look at the way they are actually distributed. Every cell harbours just one of them (or two, if we bear in mind diploidy). Since we know the power this tiny central office has in the isolated cell, do they not resemble stations of local government dispersed through the body, communicating with each other with great ease, thanks to the code that is common to all of them?

Well, this is a fantastic description, perhaps less becoming a scientist than a poet. However, it needs no poetical imagination but only clear and sober scientific reflection to recognize that we are here obviously faced with events whose regular and lawful unfolding is guided by a "mechanism" entirely different from the "probability mechanism" of physics. For it is simply a fact of observation that the guiding principle in every cell is embodied in a single atomic association existing only in one copy (or sometimes two)—and a fact of observation that it results in producing events which are a paragon of orderliness. Whether we find it astonishing or whether we find it quite plausible that a small but highly organized group of atoms be capable of acting in this manner, the situation is unprecedented, it is unknown anywhere else except in living matter. The physicist and the chemist, investigating inanimate matter, have never witnessed phenomena which they had to interpret in this way. The case did not arise and so our theory does not cover it—our beautiful statistical theory of which we were so justly proud because it allowed us to look behind the curtain, to watch the magnificent order of exact physical law coming forth from atomic and molecular disorder; because it revealed that the most important, the most general, the all-embracing law of entropy increase could be understood without a special assumption *ad hoc,* for it is nothing but molecular disorder itself.

Two Ways of Producing Orderliness

The orderliness encountered in the unfolding of life springs from a different source. It appears that there are two different "mechanisms" by which orderly events can be produced: the "statistical mechanism" which produces "order from disorder" and the new one, producing "order from order." To the unprejudiced mind the second principle appears to be much simpler, much more plausible. No doubt it is. That is why physicists were so proud to have fallen in with the other one, the "order-from-disorder" principle, which is actually followed in Nature and which alone conveys an understanding of the great line of natural events, in the first place of their irreversibility. But we cannot expect that the "laws of physics" derived from it suffice straightaway to explain the behaviour of living matter, whose most striking features are visibly based to a large extent on the "order-from-order" principle. You would not expect two entirely different mechanisms to bring about the same type of law—you would not expect your latch-key to open your neighbour's door as well.

We must therefore not be discouraged by the difficulty of interpreting life by the ordinary laws of physics. For that is just what is to be expected from the knowledge we have gained of the structure of living matter. We must be prepared to find a new type of physical law prevailing in it. Or are we to term it a non-physical, not to say a superphysical, law?

The New Principle Is Not Alien to Physics

No. I do not think that. For the new principle that is involved is a genuinely physical one: it is, in my opinion, nothing else than the principle of quantum theory over again. To explain this, we have to go to some length, including a refinement, not to say an amendment, of the assertion previously made, namely, that all physical laws are based on statistics.

This assertion, made again and again, could not fail to arouse contradiction. For, indeed, there are phenomena whose conspicuous features are visibly based directly on the "order-from-order" principle and appear to have nothing to do with statistics or molecular disorder.

The order of the solar system, the motion of the planets, is maintained for an almost indefinite time. The constellation of this moment is directly connected with the constellation at any particular moment in the times of the Pyramids; it can be traced back to it, or vice versa. Historical eclipses have been calculated and have been found in close agreement with historical records or have even in some cases served to correct the accepted chronology. These calculations do not imply any statistics, they are based solely on Newton's law of universal attraction.

Nor does the regular motion of a good clock or of any similar mechanism appear to have anything to do with statistics. In short, all purely mechanical events seem to follow distinctly and directly the "order-from-order" principle. And if we say "mechanical," the term must be taken in a wide sense. A very useful kind of clock is, as you know, based on the regular transmission of electric pulses from the power station.

I remember an interesting little paper by Max Planck on the topic "The Dynamical and the Statistical Type of Law" ("Dynamische und Statistische Gesetzmässigkeit"). The distinction is precisely the one we have here labelled as "order from order" and "order from disorder." The object of that paper was to show how the interesting statistical type of law, controlling large-scale events, is constituted from the "dynamical" laws supposed to govern the small-scale events, the interaction of the single atoms and molecules. The latter type is illustrated by large-scale mechanical phenomena, as the motion of the planets or of a clock, etc.

Thus it would appear that the "new" principle, the order-from-order principle, to which we have pointed with great solemnity as being the real clue to the understanding of life, is not at all new to physics. Planck's attitude even vindicates priority for it. We seem to arrive at the ridiculous conclusion that the clue to the understanding of life is that it is based on a pure mechanism, a "clock-work" in the sense of Planck's paper. The conclusion is not ridiculous and is, in my opinion, not entirely wrong, but it has to be taken "with a very big grain of salt."

The Motion of a Clock

Let us analyse the motion of a real clock accurately. It is not at all a purely mechanical phenomenon. A purely mechanical clock would need no spring, no winding. Once set in motion, it would go on for ever. A real clock without a spring stops after a few beats of the pendulum, its mechanical energy is turned into heat. This is an infinitely complicated atomistic process. The general picture the physicist forms of it compels him to admit that the inverse process is not entirely impossible: a springless clock might suddenly begin to move, at the expense of the heat energy of its own cog wheels and of the environment. The physicist would have to say: The clock experiences an exceptionally intense fit of Brownian movement. We have seen in chapter 2 that with a very sensitive torsional balance (electrometer or galvanometer) that sort of thing happens all the time. In the case of a clock it is, of course, infinitely unlikely.

Whether the motion of a clock is to be assigned to the dynamical or to the statistical type of lawful events (to use Planck's expressions) depends on our attitude. In calling it a dynamical phenomenon we fix attention on the regular going that can be secured by a comparatively weak spring, which overcomes the small disturbances by heat motion, so that we may disregard them. But if we remember that without a spring the clock is gradually slowed down by friction, we find that this process can only be understood as a statistical phenomenon.

However insignificant the frictional and heating effects in a clock may be from the practical point of view, there can be no doubt that the second attitude, which does not neglect them, is the more fundamental one, even when we are faced with the regular motion of a clock that is driven by a spring. For it must not be believed that the driving mechanism really does away with the statistical nature of the process. The true physical picture includes the possibility that even a regularly going clock should all at once invert its motion and, working backward, rewind its own spring—at the expense of the heat of the environment. The event is just "still a little less likely" than a "Brownian fit" of a clock without driving mechanism.

Clockwork After All Statistical

Let us now review the situation. The "simple" case we have analysed is representative of many others—in fact of all such as appear to evade the all-embracing principle of molecular statistics. Clockworks made of real physical matter (in contrast to imagination) are not true "clock-works." The element of chance may be more or less reduced, the likelihood of the clock suddenly going altogether wrong may be infinitesimal, but it always remains in the background. Even in the motion of the celestial bodies irreversible frictional and thermal influences are not wanting. Thus the rotation of the earth is slowly diminished by tidal friction, and along with this reduction the moon gradually recedes from the earth, which would not happen if the earth were a completely rigid rotating sphere.

Nevertheless the fact remains that "physical clock-works" visibly display very prominent "order-from-order" features—the type that aroused the physicist's excitement when he encountered them in the organism. It seems likely that the two cases have after all something in common. It remains to be seen what this is and what is the striking difference which makes the case of the organism after all novel and unprecedented.

Nernst's Theorem

When does a physical system—any kind of association of atoms—display "dynamical law" (in Planck's meaning) or "clock-work features"? Quantum theory has a very short answer to this question, viz. at the absolute zero of temperature. As zero temperature is approached the molecular disorder ceases to have any bearing on physical events. This fact was, by the way, not discovered by theory, but by carefully investigating chemical reactions over a wide range of temperatures and extrapolating the results to zero temperature—which cannot actually be reached. This is Walther Nernst's famous "Heat Theorem," which is sometimes, and not unduly, given the proud name of the "Third Law of Thermodynamics" (the first being the energy principle, the second the entropy principle.)

Quantum theory provides the rational foundation of Nernst's empirical law, and also enables us to estimate how closely a system must approach to the absolute zero in order to display an approximately "dynamical" behaviour. What temperature is in any particular case already practically equivalent to zero?

Now you must not believe that this always has to be a very low temperature. Indeed, Nernst's discovery was induced by the fact that even at room temperature entropy plays an astonishingly insignificant role in many chemical reactions. (Let me recall that entropy is a direct measure of molecular disorder, viz. its logarithm.)

The Pendulum Clock is Virtually at Zero Temperature

What about a pendulum clock? For a pendulum clock room temperature is practically equivalent to zero. That is the reason why it works "dynamically." It will continue to work as it does if you cool it (provided that you have removed all traces of oil!). But it does not continue to work if you heat it above room temperature, for it will eventually melt.

The Relation Between Clockwork and Organism

That seems very trivial but it does, I think, hit the cardinal point. Clockworks are capable of functioning "dynamically," because they are built of solids, which are kept in shape by London-Heitler forces, strong enough to elude the disorderly tendency of heat motion at ordinary temperature.

Now, I think, few words more are needed to disclose the point of resemblance between a clockwork and an organism. It is simply and solely that the latter also hinges upon a solid—the aperiodic crystal forming the hereditary substance, largely withdrawn from the disorder of heat motion. But please do not accuse me of calling the chromosome fibres just the "cogs of the organic machine"—at least not without a reference to the profound physical theories on which the simile is based.

For, indeed, it needs still less rhetoric to recall the fundamental difference between the two and to justify the epithets novel and unprecedented in the biological case.

The most striking features are: first, the curious distribution of the cogs in a many-celled organism, for which I may refer to the somewhat poetical description [on p. 00] and secondly, the fact that the single cog is not of coarse human make, but is the finest masterpiece ever achieved along the lines of the Lord's quantum mechanics.

NOTES

[1] Neither can the body determine the mind to think, nor the mind determine the body to motion or rest or anything else (if such there be).

[2] To state this in complete generality about "the laws of physics" is perhaps challengeable. The point will be discussed in chapter 7.

[3] If a man never contradicts himself, the reason must be that he virtually never says anything at all.

Part II

INTRODUCTION

An area of great importance to philosophers and biologists alike is that of problems relating to explanation and methodology in biological science. Similarly, physicists and philosophers share problems concerning causal explanation, probability, deductive explanation, experimental method, and the character of scientific laws. Indeed, physics serves as the paradigm example of the science in which these problems emerge with the greatest force and clarity. In biology, however, problems of philosophical interest often arise because of the *differences* which exist between biology and physics. For instance, the issue of teleological explanation and teleological laws is a very pressing one for people interested in biological explanation. In addition, the relationship which this mode of explanation bears to causal or mechanical explanation in the physical sciences demands clarification. It is not merely the intrinsic quality of the type of explanation—and I don't mean to diminish its significance—but also its relation to that mode of explanation in the physical sciences from which it differs which makes it a subject of consideration.

An excellent example of how biological explanation is afforded a heightened interest is provided in the selection by Braithwaite. Arguing against some thinkers who may choose to regard teleological explanation as ultimately improper, Braithwaite notes that it does provide intellectual satisfaction in certain spheres of consideration. (Biology and psychology are two favorite examples of most philosophers.) For Braithwaite, the relationship between teleological explanation and ordinary causal explanation is of great importance and it is this point which he investigates in considerable detail. He establishes clearly those conditions under which it is justifiable, indeed, necessary, to offer a teleological explanation rather than a causal one. He is quite insistent to point out that when intellectual satisfaction is achieved, then the goal of the individual offering the explanation has been met. To dismiss teleological explanation on any ground other than this is simply wrong.

An example of such an incorrect dismissal offered by Braithwaite illustrates the manner in which intelligent people often become confused because they are not cognizant of the level of discourse on which they are communicating. The example which Braithwaite offers has to do with those biologists who consider explanation in terms of final causes as poor explanation, because the advance of biology will soon render all biological phenomena intelligible in terms of strict causal explanation. Braithwaite characterizes this attitude as follows:

> This attitude is equivalent to an attempt to solve the problem of teleological explanations in a second way, by reducing them to physico-chemical explanations of the ordinary causal sort. It is admitted that biochemistry and biophysics at the moment cannot effect this reduction in the great majority of cases, but it is expected that some day they will be able to. Teleological explanations must be accepted as irreducible to causal explanations at present, but not as in principle irreducible. Thus the philosophical problem presented by the reference to the future in such explanations is a temporary problem only, to be solved by the progress of science. A teleological explanation is to be regarded as a very poor sort of explanation indeed, to be discarded as soon as the real, physico-chemical causes have been discovered.

Upon reflection it should become quite clear that the progress of science or the content of biochemistry have nothing whatever to do with the *structure* of a particular mode of explanation. An argument against teleological explanation should not have as its basis the belief that biology will soon be explained by physics. The latter point of view is nothing more than a consideration of the relationship which adheres between two different sciences. It says nothing about their respective modes of explanation. The confusion arises because physics uses physico-chemical explanation (which is not to be confused with pure causal explanation), while biology often employs teleological explanation. Consequently, to state that one science will ultimately be incorporated into the other is to mention only incidentally that one type of explanation, namely teleological, will not be necessary any longer. However, this is not to argue in any way that teleological explanation is a poor *type* of explanation. It is important, then, when we consider some of the problems in this section of the book, to keep track of our level of discourse. To indict teleological explanation we must attack the *structure* of that mode of explanation and not the *content* of the science which it is employed to describe. All too often, particularly in the philosophy of biology, there is profound confusion between problems of ontology and explanation. If that confusion can be avoided, then a great step toward the clarification of certain biological and philosophical issues will have been taken.

Before continuing, it is important to note that Braithwaite mentions a criterion for causal laws which is stated in the previous chapter of his book. There are four criteria in all, and I think that the present selection will be better understood if the following passage from Braithwaite is included here:

Let us leave aside for the moment the regular concomitances, and consider the case in which the natural law asserts a constant conjunction of properties in two distinct events. There are several plausible criteria with claims to be used for selecting, out of the class of natural laws of this two-event sort, a subclass of causal laws. Here are some of the most plausible ones:

I. The weakest criterion is perhaps to exclude only natural laws asserting regular precedences and to admit those asserting regular simultaneities as well as regular sequences.

II. A stronger criterion would admit only natural laws asserting regular sequences, but would admit all these whatever the interval of time between the event having the property A (the "cause-event").

III. A yet stronger criterion would admit only those natural laws asserting regular sequences where these regular sequences were deducible in an established deductive system from regular sequences in which there was no time-interval between the cause-event and the effect-event, i.e. in which the cause-event and the effect-event were temporally continuous.

IV. An even stronger criterion could admit only these temporally continuous regular sequences.

The selection by Bertalanffy, from one of his most important works, is rather self-explanatory. It is Bertalanffy's contention that 20th century biology is in a critical state because it has yet to establish a clear theoretical foundation upon which the facts which constitute the building blocks of the science can be placed. Without this theoretical framework there is no science; hence, a crisis emerges. The reason for its present appearance is that biological investigation is proceeding at such a rapid rate that it is becoming increasingly important to introduce a theoretical structure which will serve to unify and arrange the results of this activity. Just as physics requires a theoretical consideration of itself, so biology calls for the same. Bertalanffy sums up this point of view quite clearly when he says:

We see that biological knowledge operates at three levels: in the first level it deals with the ordering, the simple and comparative description, of its objects. In the second the causal, organismic, and historical connexions of the organism are investigated, and—in "general biology"—rules are set up for the uniformities which here present themselves. The third stage—that of theoretical biology—yields with the help of hypothetical suppositions, the laws of biological processes.

It is the view of Bertalanffy that biology has not yet gone through its own Copernican revolution. Needless to say, then, its Einstein has yet to emerge. Theoretical biology must arise as a full-blown area of consideration in order for biology to begin to approach the level of physics. It must operate in two ways. In the first place, it must play a critical role by investigating various biological theories and determining where their strengths and shortcomings lie. This critical function is a continuous and essential one for theoretical biology in that it serves to prevent biologists from being led astray through improper analyses of the material available to them. The second role of theoretical biology is constructive, in that it attempts to establish a unitary system of biological thought along the lines of a hypothetico-deductive model. Bertalanffy succeeds in conveying to his reader a sense of the great need which biology has for a theoretical superstructure.

There is also another ground for the anti-theoretical attitude of contemporary biologists which is not difficult to understand and agree with. Only too often do we see the theorist leave the solid ground of experience and experiment and disappear into the blue mists of metaphysical speculation. When once the aversion to *this* kind of theoretical biology has seized biologists, it may easily happen that *every* kind of "theory" comes to be regarded as a departure from his proper scientific business. Here, then, is another point where there is a need for change in contemporary attitude, a change which ought not to consist in the rejection of theory in general but in taking seriously the need for a scientific theoretical biology, whilst at the same time declaring war upon all such light-minded speculation as has been responsible for the mistrust of "theory" in biology.

Joseph Woodger provides an example of the nature of theoretical biology by presenting a theoretical model of the organism. It is Woodger's position that biology ought to have a theoretical foundation in much the same way that physics does. Since the publication of his article, theoretical biology has begun to make its appearance, but is certainly an embryonic science with respect to many of its colleagues within the biological domain. The failure of biology to develop a clear theoretical base lies, in Woodger's opinion, in the fact that biology has not adopted the rigorous method of mathematics. It is not that biology must necessarily become mathematical in order to be theoretical; it is only that the rigor and procedural exactitude of mathematics must be adopted by individuals who seek to investigate the theoretical foundations of biology.

Woodger deals on two levels in this paper. On one level he speaks of methodological procedures which biologists employ. On another level, he discusses the various problems of fruitful explanation. Quite often the two levels are mixed together in the same discussion, and it is rare that such a clear distinction between the two is drawn. The reader will notice that Woodger is very careful to point out just what he is talking about and how he is carrying out his analysis. He takes the familiar philosophic point of view that a mere understanding of biological statements is not sufficient for the development of a theoretical biology. We must be sure, says Woodger, of two things: first, we must understand the foundations of our methodological procedures and second, we must have a clear understanding of the epistemological position we are adopting as biological theoreticians. With respect to the first point, he says, ". . . so long as the investigator of nature continues to make discoveries (and the volume of papers issued in the quarterly journals shows no sign of decline) there would seem to be no occasion to deflect attention from the business of investigating nature to matters relating to the process of investigation itself." With regard to the second point he states:

> Now it is clear that the sort of systematized knowledge about the instruments and dyes, etc., which the investigator uses to guide him in his pursuits, is natural scientific knowledge. It is applied physics and chemistry. But the sort of knowledge required for the guidance of the process of interpretation will be knowledge about the properties of knowledge itself, and will not be natural scientific knowledge. It seems to be the case as a general rule that the people who pursue natural scientific knowledge never pay much attention to knowledge about knowledge itself, and the people who make knowledge itself an object of scientific investigation do not always know much about the subject matter of natural scientific knowledge. It is this state of affairs which is responsible for a situation comparable to that of the cook and the baking powder. Everyone recognizes the desirability of knowing something about the physical instruments used in scientific investigation, but the importance of understanding the properties of the intellectual tools involved—concepts, propositions, principles of inference, "working hypotheses," postulates, etc.,—is much less clearly appreciated. And from the standpoint of the process of interpretation this may be a misfortune.

In the above passage, Woodger points out a fascinating and important point. All too often in the philosophy of science people engage in critical analyses of various scientific concepts without having a detailed understanding of the science they are speaking of. The result is generally a rather sterile analysis which suffers from the condition of distance. That is, in order to have meaningful insights into any particular science it is generally necessary to be somewhat conversant in the language of that science and even more, to have dealt first hand with the materials and problems with which that discipline is concerned. On the other hand, it is all too often the case that scientists, in attempting an analysis of what their work is all about, produce very little fruitful material. The problem here is that little is known by these people of the power of philosophical analysis and their reflections often drift off into the realm of opinion which may be engaging, but does not suffice to serve as a thoroughgoing analysis of a particular scientific enterprise.

In the main section of the paper, Woodger offers an example of what a strict philosophical analysis of the concept of organism might be like. He uses the philosophical tool of symbolic logic and constructs a view of the living system. His major point is that an understanding of the organism requires an understanding of the various organizational relations which obtain between different parts of the total system. It is not that living things are made of any special substance not found in non-living things; instead, it is that the peculiar nature of these entities which we call living is to be seen in the manner in which they are put together. Woodger introduces the notion of hierarchial organization into the picture he draws of the living system. By exploiting the meaning of this concept he presents a fascinating view of the theoritical structure of the organism. His particular approach is not unlike that of Bertalanffy, except for the fact that Woodger employs symbolic logic as a large part of all his work. It is his opinion that the use of this tool lends clarity to what he says. While this view is not held by all philosophers of science, it is certainly an important element in contemporary philosophical analysis, and it is left for the reader to decide how valuable the approach is for him.

The final article in this section, written by Ernest Nagel, is concerned with three different but related dimensions of organismic biology. Nagel treats the latter as a theory, a mode of explanation, and a methodology. Perhaps the greatest value of his article in this context is that it is a model of clear analysis. Rather than taking the position of organismic biology, Nagel seeks to explicate and compare. The chief opponent to the theory is, of course, mechanism, and the antagonism between the two points of view concerning the nature of life is lucidly presented. Unlike Bertalanffy and Woodger, upon both of whom he draws a great deal for his discussion, Nagel remains neutral in his discussion. He thus serves to bring together all of the various pros and cons of organismic biology as well as some of the points expressed by the mechanists. In general, the selection requires little additional explication in the light of the other three articles in this selection, and the reader is invited to enjoy whatever insights each may have to offer.

R. B. Braithwaite

Richard Beran Braithwaite was born in 1900. His advanced education was at King's College in Cambridge where he became a Fellow in 1924. He was president of the Mind Association in 1946 and the Aristotelian Society from 1946 to 1947. In 1950 he was Annual Philosophical Lecturer to the British Academy and from 1961 to 1963 was president of the British Society for the Philosophy of Science. Among Braithwaite's publications are: *Scientific Explanation, Theory of Games as a Tool for the Moral Philosopher,* and *An Empiricist's View of the Nature of Religious Belief.* Currently, he is living in Cambridge and is Emeritus Professor of Moral Philosophy at the University of Cambridge.

The following selection is taken from *Scientific Explanation,* (1953), pp. 319-41, and is reprinted here with the kind permission of Cambridge University Press.

SCIENTIFIC EXPLANATION

Chapter X
Causal and Teleological Explanation

Any proper answer to a "Why?" question may be said to be an explanation of a sort. So the different kinds of explanation can best be appreciated by considering the different sorts of answers that are appropriate to the same or to different "Why?" questions.

What is demanded in a "Why?" question is intellectual satisfaction of one kind or another, and this can be provided, partially or completely, in different ways. Frequently the questioner does not know beforehand what sort of answer will satisfy him. And what gives partial or complete intellectual satisfaction to one person may give none whatever to a person at a different stage of intellectual development. A small child, for example, is frequently satisfied by a confident reassertion of the fact about which he has asked the "Why?" question. This is not foolishness on his part. The child is prepared to accept the fact without question on authority; what he is doubtful about is whether the authority is good enough, and a confident reassertion by the person to whom he has asked the "Why?" question may serve to strengthen the authority sufficiently for him to fell complete satisfaction in accepting it.

When an adult wishes for satisfaction of this purely confirmatory sort, he phrases his question in the form "Is it really the case that . . . ?" reserving his "Why?" questions for cases in which he requires for satisfaction something more than a repetition of the "Why?" sentence with the omission of the "Why." What he requires is explanation in the proper sense of the proffering of explanans—proposition as an explanation of the explicandum-fact about which he has asked the "Why?" question, the explicans being required to be different from the explicandum.

Different sorts of explicanda call for different sorts of explanation— or, at least, for different sorts of first-stage explanations. The primary explicanda for science are particular empirical facts; and first-stage explanations of these are of two types. When an adult asks "Why f?" of a particular matter of fact f, he is usually wanting either a *causal explanation* expressed by the sentence "Because of g" or a *teleological explanation* expressed by the sentence "In order that g." Each of these types of explanation will involve an explicit or implicit reference to scientific laws; a "Why?" question asked of a scientific law (and this may well be a second-stage "Why?" question asked of a particular matter of fact) will be a request, not for a cause or for a teleological goal, but for a reason for the scientific law being what it is. This chapter will be devoted to the first-stage explanation of particular empirical facts, a discussion of the explanation of scientific laws themselves being postponed until the next chapter.

Causal Explanation

Our discussion in the last chapter of the meaning of the sentence "q because p" covers most that need be said of causal explanation. When a person asks for a cause of a particular event (e.g. the fall of this picture to the floor at noon yesterday), what he is requesting is the specification of a preceding or simultaneous event which, in conjunction with certain unspecified cause factors of the nature of permanent conditions, is nomically sufficient to determine the occurrence of the event to be explained (the explicandum-event) in accordance with a causal law, in one of the customary senses of "causal law."[1] The "Why?" question is not expected to be answered by detailing all the events which together make up a total cause, i.e. a set of events which collectively determine the explicandum-event; all that is usually expected is the part-cause which is of most interest to the questioner—which presumably is that of which he is ignorant. One sense of giving a *complete* explanation would be that of specifying a total cause; in this sense, as indeed in most senses of complete explanation, a "complete explanation" will not be unique, since (in almost all sense of "cause") the same event can perfectly well have many different total causes.

There are various complications about causal explanations considered as answers to "Why?" questions which need not detain us long. The formal explanation just given is that in which the questioner is taken to be asking for a *sufficient condition* for the explicandum-event, or for part of a sufficient condition, the other part being supposed to be already known. The explicans in such an explanation is an event the occurrence of which possessing a certain property, in conjunction with other events with suitable properties, nomically determines the occurrence of the explicandum-event with a certain property. So the existence of the explicans event ensures the existence of the explicandum-event. But the "Why?" question is sometimes a request for a *necessary condition* for the explicandum-event; it then asks for the specification of an event which is such that, had it not occurred, the explicandum-event would also not have occurred. In this case it is the explicans-event which is nomically determined by the explicandum-event instead of the other way around. And frequently the "Why?" question requires as answer the specification of an event which is both one of a set of events which together form a sufficient condition and one which in the presence of the rest of the set of events is a necessary condition for the occurrence of the explicandum-event.

There is one type of causal explanation which, rightly or wrongly, gives great intellectual satisfaction to those who have been educated in the contemporary natural sciences—namely, causal explanations making use of causal laws which are causal according to criterion III (p.137). Here an event to be acceptable as explicans must be the first member of a *causal chain* of events ending with the explicandum, a spatio-temporally continuous chain of events being said to form a causal chain if every event in the chain nomically determines its neighbours in the chain in such a way that the causal law relating the explicans-event with the explicandum-event is a consequence within a true deductive system of higher-level laws which relate only spatio-temporally continuous events. The intellectual satisfaction provided by an explanation which cites a "causal ancestor" is due partly to the great success of such explanations in the physical sciences, but partly also to the fact that, if the deductive system whose highest-level hypotheses relate spatio-temporally continuous events is unrefuted by the evidence, there will be a great deal of evidence which supports it. For an unlimited number of lower-level hypotheses about causal ancestries will be deducible from the highest-level hypotheses, and the falsity of some, at least, of these would be expected to leap to the eye if the highest-level hypotheses were false.

At the other extreme there is a type of explanation based upon generalizations which have been established by direct induction without any indirect hypothetico-deductive support. These generalizations, therefore, do not satisfy the conditions for being entitled "natural laws," and consequently cannot be classed as causal laws by even the weakest criterion of the last chapter (criterion I). So these explanations cannot be called causal explanations. Nevertheless they can give some intellectual satisfaction, for they give information on one point about which the questioner may be ignorant. Molière was right in laughing at the doctors who offered the *virtus dormitiva* of opium as the answer to the question as to why opium produced sleep.[2] But it would not be foolish to answer the question as to why a particular specimen of powder produced sleep by replying that it was because that powder was opium, and that opium had the property of producing sleep. For this would inform a questioner who did not know it already that the powder produced sleep, not by virtue of its colour, degree of powderedness, etc., but by virtue of its chemical composition. Similarly, when a child asks "Why is this bird white?" the reply "Because it is a swan, and all English swans are white" tells him that the whiteness is not a peculiarity of this particular bird, and thus shows the particular case to be an instance of a general proposition.

Teleological Explanation

We must now turn to a type of explanation which has so far not been discussed—a type which has given rise to a great deal of discussion among philosophers and philsophically minded biologists, because it has been thought to raise peculiar scientific and philosophical difficulties. This type of explanation is that in which the "Why?" question about a particular event or activity is answered by specifying a goal or end towards the attainment of which the event or activity is a means. Such explanations will be called "Teleological explanations."[3] If I am asked why I am staying in Cambridge all through August, I should reply "In order to finish writing my book"; to reply thus would be to give a teleological explanation. If I am asked why my cat paws at the door on a particular occasion, I might well reply "In order that I should open the door for it"—another teleological explanation. If an ornithologist is asked why a cuckoo lays its egg in the nest of another bird, and replies "So that the other bird may hatch out and nurture its young," or if a physiologist is asked why the heart beats, and replies "To circulate the blood round the body" or (in more detail) "To convey oxygen from the lungs to the tissues and carbon dioxide from the tissues to the lungs" or (in terms of an ultimate biological end) "In order that the body may continue to live," he will be giving in each case a teleological explanation of the action in terms of the goal or end of the action. The explanation consists in stating a goal to be attained: it describes the action as one directed towards a certain goal—as a "goal-directed activity" (to use E. S. Russell's convenient phrase[4]), the word "directed" being used (as it will here be used) to imply a direction but not to imply a director.

If we take an explanation (as we are doing) to be any answer to a "Why?" question which in any way answers the question, and thereby gives some degree of intellectual satisfaction to the questioner, there can be no doubt that teleological answers of the sort of which I have given examples are genuine explanations. The fact that they all may give rise to further questions does not imply that they are not perfectly proper answers to the questions asked. My answer as to why I am staying in Cambridge all through August would almost certanly not lead to a further question, unless my friend wished to start a philosophical discussion as to the correct analysis of the motives of rational action. My answer as to why my cat paws the door might lead to the further question as to why the cat (to use common-sense language) "wants to be let out," to which another teleological answer would be appropriate, or to the question as to how the cat has learnt to paw the door to show that he wants to be let out, which would lead to a description, which might or might not be in teleological terms, of the processes of learning in cats. But all these would be regarded as further and different questions; the first simple teleological answer would be taken as what the questioner was asking for, and if it did not give him adequate intellectual satisfaction, he would expect not to repeat the question but to ask another.

But, having insisted that teleological explanations are perfectly good first-stage explanations, we have to admit that they have one feature which distinguishes them from causal explanations, and that this feature has proved very puzzling to philosophers, whether concerned with philosophical psychology or with the philosophy of biology. In a causal explanation the explicandum is explained in terms of a cause which either precedes or is simultaneous with it: in a teleological explanation the explicandum is explained as being causally related either to a particular goal in the future or to a biological end which is as much future as present or past. It is the reference in teleological explanations to states of affairs in the future, and often in the comparatively distant future, which has been a philosophical problem ever since Aristotle introduced the notion of "final cause"; the controversy as to the legitimacy of explanations in terms of final causes rages continually among philosophers of biology and, to a less extent, among working biologists.

Now there is one type of teleological explanation in which the reference to the future presents no difficulty, namely, explanations of an intentional human action in terms of a goal to the attainment of which the action is a means. For my teleological answer to the question as to why I am staying in Cambridge all through August—that I am doing so in order to finish writing my book—would be regarded by my questioner as equivalent to an answer that I am doing so because I intend to finish writing my book, my staying in Cambridge being a means to fulfil that intention; and this answer would have been an explanation of the causal sort with my intention as cause preceding my stay in Cambridge as effect.[5] Teleological explanations of intentional goal-directed activities are always understood as reducible to causal explanations with intentions as causes; to use the Aristotelian terms, the idea of the "final cause" functions as "efficient cause"; the goal-directed behaviour is explained as goal-intended behaviour.

This is not to say that there is no philosophical difficulty about intentional action; there is the problem—fundamental for philosophical psychology—as to the correct analysis of the intention to act in a certain way. But this is different from our problem as to how a future reference can occur in an explanation, unless indeed an extreme behaviouristic analysis is adopted, according to which there is no conscious element in an intention, and goal-intended behaviour is simply what we call goal-directed behaviour in the higher animals. But for this extreme behaviourism psychology reduces to biology, and intentional action falls under biological goal-directed activity and the type of teleological explanation we meet in the sciences concerned with life in general and not especially with mind.

The difficulty about the future reference occurs then in all teleological explanations which are not reducible to explanations in terms of a conscious intention to attain the goal. Here one cannot obviously reduce the teleological answer, which explains a present event by means of a future event, to a non-teleological answer in terms of a present or past cause. It is teleological explanations which cannot obviously be so reduced which present the philosophical problem; and the rest of this chapter will be devoted to this type of teleological explanations and to the problems raised by them.

Current Attempts to Eliminate Final Causes

There are two ways of solving this problem which are fashionable today. Both are attempts to reduce all teleological explanations to causal explanations, and thus to eliminate the special puzzle presented by future reference; but the attempts are made in opposite directions.

The first way is to emphasize the similarity between teleological explanations of the type with which we are now concerned and the teleological explanations of intentional actions in which the future reference can be explained away, and to argue by analogy that in all cases the teleological explanation is reducible to one in which an intention, or something analogous to an intention, in the agent is the "efficient cause," so that goal-directed activity is always a sort of goal-intended activity. My cat's behaviour in pawing at the closed door, it may be said, is sufficiently similar to a man's behaviour in knocking at a locked door for it to be reasonable to infer that the cat, like the man, is acting as it does because of a conscious intention, or at least a conscious desire, to be let through the door. Similarly a neurotic's goal-directed behaviour may be explained by his having an unconscious intention or desire; a bird's nest-building by its having an instinct to do so. When the goal-directed activity to be explained is that of a part of a whole organism, as in my example of the heart's beating, the analogue to the intention—the drive or conatus or nisus or urge—is usually posited not in the separate organ but in the organism as a whole—an urge towards self-preservation, for example. Sometimes the analogy is pressed so far that a purposiveness similar to that of voluntary action is assumed in all teleological behaviour. William McDougall, for instance, after explaining that by "purposiveness" in human movements he means not only that "they are made for the sake of attaining their natural end" (i.e. that they are teleological in my sense), but that "this end is more or less clearly anticipated or foreseen," goes on to speak of a "scale or degrees of purposiveness," at the lower end of which there is a "vague anticipation of the goal" which may also be ascribed to an animal's goal-directed behaviour.[6]

Other writers (e.g. E. S. Russell) would reject as unduly anthropomorphic the attribution of purposiveness to such activities, and would describe the efficient cause as a conatus or drive. But all writers who deal with the problem of teleological explanation in the first way agree in postulating something in the organism which is present whenever goal-directed behaviour is taking place and which is to explain it in the ordinary causal way, and agree in supposing that this something cannot be analysed purely in physico-chemical terms.

The biological orthodoxy of to-day, however, would say that the postulation of this "something," not explicable in physico-chemical terms, to account for teleological behaviour was an assumption which was either methodologically vicious (if the "something" was supposed to have no properties other than that of being the cause of the goal-directed behaviour) or metaphysical and non-empirical (if it was supposed to have additional properties such as McDougall's purposiveness). And orthodox biologists would go on to say that satisfactory explanations had been given of many goal-directed activities in physico-chemical terms, and that as the new sciences of biochemistry and biophysics advance, there is less and less reason to suppose that there will be any teleological action (or at any rate any teleological action in which consciousness is not involved) that will not be explicable by means of the concepts and laws of chemistry and physics alone.

This attitude is equivalent to an attempt to solve the problem of teleological explanations in a second way, by reducing them to physico-chemical explanations of the ordinary causal sort. It is admitted that biochemistry and biophysics at the moment cannot effect this reduction in the great majority of cases, but it is expected that some day they will be able to. Teleological explanations must be accepted as irreducible to causal explanations at present, but not as in principle irreducible. Thus the philosophical problem presented by the reference to the future in such explanations is a temporary problem only, to be solved by the progress of science. A teleological explanation is to be regarded as a very poor sort of explanation indeed, to be discarded as soon as the real, physico-chemical causes have been discovered.

It seems to me that the orthodox biologists are right in rejecting the postulation of a conatus or drive which is non-physical and *sui generis* in order to explain the goal-directed behaviour which they meet in their biological studies, but wrong in minimizing the intellectual satisfaction to be derived from teleological explanations. I believe that we can go on the orthodox assumption that every biological event is physico-chemically determined, and yet find an important place in biology for such explanations. So what I propose to do is to try to give an account of the nature of teleological explanations which will resolve the philosophical difficulty about the apparent determination of the present by the future without either contravening the usual determination principles of science or reducing all biological laws to those of chemistry and physics.[7]

Teleological Causal Chains

If we make the ordinary determination assumptions of physical science, the apparent determination of the explicandum by a future event in a teleological explanation is not direct, but works by means of a causal chain of events lying between the explicandum and the goal. Even in intentional action the intention does not directly produce the goal: it starts a chain of action whose final stage is attainment of the goal. In non-intentional goal-directed action the goal-directedness consists simply in the fact that the causal chain in the organism goes in the direction of the goal, unless one wishes to suppose that there is always an extra "something"—conatus or drive—involved in goal-directedness, an assumption which I do not wish to make. Thus the notion of causal chain is fundamental. Of course this notion is equally fundamental in the non-teleological explanations provided by the physical sciences, where the explaining cause is frequently given not as a preceding event connected with the explicandum by a causal chain. Let us approach our problem, therefore, by asking what (if any) is the peculiarity of the causal chains which are involved in teleological explanations.

Bertrand Russell, in his behaviouristic account of desire, approached our problem in the same way as I am doing by asserting that the peculiarity of teleological causal chains of actions is that they form "behaviour-cycles." But the only criterion he gave to enable us to pick out the behaviour-cycles from other repeated series of events in the life of an animal was that the final stage in a behaviour-cycle is "normally a condition of temporary quiescence."[8] He illustrated this by an animal falling asleep after it has eaten. But temporary quiescence is quite inadequate to serve as the *differentia* for which we are seeking. After a bomb has exploded, or a volcano ceased to erupt, a state of temporary quiescence is attained. Here no teleology is concerned in the causal chains. It seems impossible to find any characteristic of the final state by itself of a teleological causal chain which is general enough to cover all the goals of goal-directed actions and yet specific enough to differentiate such actions from other repeated cycles of behaviour. It is necessary, I think, to look at the whole causal chain and not merely at its final state.

It seems to me that a distinguishing criterion can be found in one of the characteristics which biologists have emphasized in their descriptions of goal-directed behaviour, namely persistence towards the goal under varying conditions. To quote E. S. Russell: "Coming to a definite end or terminus is not *per se* distinctive of directive activity, for inorganic processes also move towards a natural terminus. . . . What *is* distinctive is the active persistence of directive activity towards its goal, the use of alternative means towards the same end, the achievement of results in the face of difficulties."[9] Examples of the "plasticity" of goal-directed behaviour will spring to every mind. To give one example only, Lashley's rats who had learnt to obtain their food by running his maze were still able to traverse the maze without false turns in order to obtain food after their powers of motor coordination had been seriously reduced by cerebellar operations, so that they could no longer run but could only crawl or lunge.[10] Plasticity is not in general a property of one teleological causal chain alone: it is a property of the organism with respect to a certain goal, namely that the organism can attain the same goal under different circumstances by alternative forms of activity making use frequently of different causal chains. Let us try to elucidate the logical and epistemological significance of this plasticity, in order to see whether it will serve our purpose of preserving the importance of teleological explanations without introducing extra-physical causation.

Consider a chain of events in a system *b*. The system may be a physical system of more or less complexity (a pilotless plane or an electron) or it may be an organic system (a complete organism or a relatively isolable part of a complete organism, e.g. the kidneys). Make the ordinary determination assumption that every event in the system is nomically determined by the whole previous state of the system together with the causally relevant factors in the system's environment or field (which will be called the "field-conditions"). Then the causal chain c of events in b throughout a period of time is nomically determined by the initial state e of the system together with the totality of field-conditions which affect the system with respect to the events in question during the period. Call this set of field-conditions f. Then, for a given system b with initial state e, c is a one-valued function of f; i.e. for given b and e, the causal chain c is uniquely determined by f—the set of field-conditions.

Now consider the property which a causal chain in a system may possess of ending in an event of type Γ without containing any other event of this type. Call this property the Γ-goal-attaining property, and the class of all causal chains having this property the Γ-goal-attaining class y. Every causal chain which is a member of y contains one and only one event of type Γ, and contains this as its final event.

Define the variancy ϕ with respect to a given system b with given initial state e, and to a given type of goal Γ, as the class of those sets of field-conditions which are such that every causal chain in b starting with e and determined by one of these sets is Γ-goal-attaining. To express this more shortly with the symbols already used, the variancy ϕ is defined as the class of those f's which uniquely determine those c's which are members of y. According to this definition, to say that a causal chain c in a system b starting from a state e ends (without having previously passed through) a state of type Γ is logically equivalent to saying that the set of field-conditions is a member of ϕ. The variancy is thus (to repeat the definition in a looser form) the range of circumstances under which the system attains the goal.

The variancy ϕ defined in relation to b, e and Γ may have no members, in which case there is no nomically possible chain in b starting from e and attaining a goal of type Γ. Or ϕ may have one member, in which case there is exactly one such chain which is nomically possible. Here the system starting from e has no plasticity; there is only one set of field-conditions which, together with e, is nomically sufficient for the attainment of a goal of type Γ

The case in which we are interested in which the system has plasticity, occurs when the variancy ϕ has more than one member, so that the occurrence of any one of alternative sets of field-conditions is together with e sufficient for the attainment of a goal of type Γ. It is important to notice that the variancy may have many members and yet there be only one nomically possible chain: it is because the size of the variancy may be greater than the number of possible causal chains that the notion of the variancy has been introduced. For it may be the case that there are various sets of field-conditions each of which, together with e, determines exactly the same causal chain. This might happen if the ultimate causal laws concerned are such that each of the events in the chain might be determined by two or more alternative field-conditions. But it more frequently happens when the events in the chain are taken as being events which attribute properties to the system as a whole, and when, although alternative field-condtions determine different part-events in the system or in parts of the system, these part-events are causally so connected that the whole event determined by them remains unchanged. For example, if the causal chain of events with which we are concerned is the chain of body temperatures throughout a period of time of one of the higher animals, a change in the relevant environmental conditions (e.g. external temperature and available sources of food) will produce changes in the activities of the animal (both changes in its total behaviour, e.g. its feeding and migration habits, and changes in its parts, e.g. its sweat glands), yet these changes will be such as to compensate for the changed environmental conditions so that the animal's body temperature does not vary. Another example would be the path of a pilotless plane, in which the machine is fitted with "feed-back" devices so designed that the plane will maintain a straight course at the correct height to the desired goal irrespective of the weather conditions it may encounter.

But usually when the variancy has more than one member, there is more than one nomically possible chain in the system in question which attains the required goal. An animal can move to get its food in many ways, a great variety of physiological processes can be called into play to repair damaged tissue, a bird can adapt its nest-building to the kind of material available. Nevertheless the essential feature, as I see it, about plasticity of behaviour is that the goal can be attained under a variety of circumstances, not that it can be attained by a variety of means. So it is the size of the variancy rather than the number of possible causal chains that is significant in analysing teleological explanation.

Let us now take the standpoint of epistemological or inductive logic and consider what are the types of situation in which we reasonably infer that there will be a goal-attaining chain of events in a system. To predict that, starting from an initial state e of a system b, there will be a causal chain which will attain a goal of type Γ is, by the definition of variancy, equivalent to predicting that the set of field-conditions which will occur will be a member of the variancy ϕ. So the reasonableness of the prediction depends upon the reasonableness of believing that ϕ is large enough to contain every set of field-conditions that is at all likely to occur.[11] Call the class of these sets of field-conditions ψ. For simplicity's sake I shall for the moment assume that we know that any set of field-conditions that will occur will be contained in ψ; that is, that the system will not in fact encounter a very unlikely environment (e.g. the next Ice Age starting suddenly to-morrow). Then the reasonableness of the prediction that the system will attain the goal depends upon the reasonableness of believing that ψ is included in ϕ.

Now there are two ways in which we may have derived our knowledge of the variancy ϕ. We may have deduced ϕ from knowledge of the relevant causal laws, or we may have inferred it inductively from knowledge of the sets of field-conditions under which similar causal chains had attained their goals in the past. In the first case, that in which the members of ϕ have been obtained by deduction, there are two interesting subcases in which we take positive steps to secure that ψ — the class of the sets of field-conditions at all likely to occur—is included in the variancy ϕ. The first subcase is that in which ϕ is small, but we deliberately arrange that ψ shall be smaller still. This happens when scientific demonstrations are performed for students in a laboratory, when elaborate precautions are taken (e.g. the experiment is done in a vacuum or distilled water is used) in order to eliminate unwanted relevant causal factors (air-currents or chemical impurities) and thus to secure that every set of conditions that may occur will fall within the known variancy ϕ so that the demonstration will be a success. The second subcase is that in which ψ is large, but we deliberately arrange that ϕ shall be larger still. This happens when a machine is deliberately designed to work under a large variety of conditions. This object may be achieved by using suitable materials: a motor-car is built to stand up to a lot of rough and careless treatment. Or it may be achieved by incorporating in the machine special self-regulating devices to ensure that the machine adjusts its method of working according to the conditions it encounters, as in the pilotless plane.

When our knowledge of the relevant variancy has been obtained by deduction from previous knowledge of the causal laws concerned, a teleological explanation of an event in terms of its goal-directedness is felt to be almost valueless.[12] For in this case, that the causal chain which will occur will lead to the goal—the "teleology" of the system—has been calculated from its "mechanism." To give a teleological answer to the "Why?" question would require forming (and suppressing) an ordinary causal answer, which would (if expressed) have given intellectual satisfaction to the questioner, in order to deduce from it a teleological answer. This would be an unprofitable, and indeed disingenuous, way of answering his question.

The situation is entirely different when our knowledge or reasonable belief about the variancy   has not been derived from knowledge of the causal laws concerned. In this case our knowledge as to what sets of conditions make up the variancy has been obtained either directly by induction from previous experience of goal-attaining behaviour that was similar to the behaviour with which we are concerned, or indirectly by deduction from general teleological propositions which have themselves been established by induction from past experience. Neither of these ways makes use of laws about the mechanisms of the causal chains. The variancy ϕ is inferred—inductively inferred—from knowledge of classes similar to ψ; that is, from past observation of the conditions under which similar teleological behaviour has taken place. For example, my knowledge of the conditions under which a swallow will migrate is derived from knowledge about past migrations of swallows and of other migrants, fortified perhaps by general teleological propositions which I accept about the external conditions for self-preservation or the survival of the species, themselves derived inductively from past experience.

It is when our knowledge of the relevant variancy has been obtained independently of any knowledge of the causal laws concerned that a teleological explanation is valuable. For in this case we are unable, through ignorance of the causal laws, to infer the future behaviour of the system from our knowledge of the causal laws; but we are able to make such an inference from knowledge of how similar systems have behaved in the past.

It should be noted that in all cases of teleological explanation of a present event by a future event, whether reducible or irreducible, inductive inferences occur at two stages of the argument. One stage is in the inference of the variancy, whether this itself is obtained inductively, or whether it is obtained deductively from causal laws or teleological generalizations which have themselves been established inductively. The other inductive stage in the argument is the inference that the set of relevant conditions that will in fact occur in the future will fall within the variancy. Every teleological answer, however reasonable, may be mistaken in each of these two ways.

But in general irreducible teleological explanations are no less worthy of credence than ordinary causal explanations. A teleological explanation of a particular event is intellectually valuable if it cannot be deduced from known causal laws: other things being equal, it is the more valuable the wider the variancy of the conditions, and hence the greater the plasticity of the behaviour concerned. It is because we are acquainted with systems—organisms and parts of organisms—which exhibit great plasticity that we make use of teleological explanations. Such an explanation may be regarded as merely another way of stating the fact of the plasticity of the goal-directed behaviour. But to state this fact is to bring the explicandum under a general category; moreover it enables us to make reliable predictions as to how the system will behave in the future. It seems ridiculous to deny the title of explanation to a statement which performs both of the functions characteristic of scientific explanations— of enabling us to appreciate connexions and to predict the future.

The analysis which has here been given of teleological explanation of non-intentional goal-directed activities supposes that the goal to which the activity is directed is later in time than the action (this indeed creates the philosophical problem) and makes great use of the notion of causal chain. It has been objected that this analysis will not cover the case of explanations of biological facts which are given in terms, not of a future goal, but of a biological end which is as much present as future; but that this case, as well as that in which the explanation is in terms of a future goal, will be covered by the more general notion of *functional explanation* in which the explanation is in terms of another part of a whole of which the explicandum is a part.[13] But the questions which seem to call for a more general functional explanation rather than for a causal-chain teleological explanation turn out on examination to be ambiguous questions. If a physiologist is asked why the heart beats, he may take this question as a request for an explanation of a particular fact, the beating of a particular heart on a particular occasion, in which case the explanation "In order to circulate the blood round the body" will also refer to the movement of the blood in a particular body on a particular occasion. But, on this interpretation, the particular movement of blood outside the heart due to a particular beating of the heart is an event whose beginning is later in time than the beginning of the event which is the heart's beating: the latter event is connected with the former event by a causal chain of events, and is a teleological explanation of it in terms of a future goal. But the physiologist may more naturally take the question, not to be a question about one particular heart on one particular occasion, but to be a question about all beatings of all hearts (or of all human hearts, or of all mammalian hearts, or etc.), in which case the question is a request for a teleological generalization of which the particular teleological explanation of the beating of one particular heart on one particular occasion would be an instance. In both cases, however, the explanation would be in terms of goal-directed activities with future goals. The peculiarity of a biological end is that it is a permanent goal; at all times during the life of the organism there are activities of the organism to be explained in terms of the biological end. My heart's beating at one moment is responsible for the circulation of my blood a short time afterwards; and my heart will have to *continue* beating for my blood to *continue* circulating. The teleological generalization of which the particular teleological explanation of the beating of my heart on a particular occasion is an instance will have instances at every moment of my life, unlike teleological laws concerning goals upon the attainment of which the animal sinks into a "temporary quiescence."[14]

Teleological Laws

We have referred in several places to teleological generalizations. Just as particular causal explanations are instances of causal propositions, of more or less generality, so particular teleological explanations are instances of teleological generalizations of more or less generality; that is, they (if true) are instances of laws according to which an event of a certain sort in a system of a certain sort is nomically determined by a later event of a certain sort in the same system. Such a teleological law will be valuable as an explanation if it has not been deduced from non-teleological laws, and it will be the more valuable both intellectually and predictively the wider the range of the variancy associated with it.

The special philosophical difficulty about teleological, as contrasted with causal, explanations of particular events—namely, that in them the present appears to be determined by the future—does not arise in the case of teleological as contrasted with non-teleological laws, considered as laws of nature without regard to their applications to yield particular explanations. For many non-teleological laws of nature, e.g. Newton's laws of mechanics, are symmetrical with respect to the earlier and later times occurring in the laws: they state that the present is determined by the future just as much as it is determined by the past. Nor do teleological laws in general differ from non-teleological ones in having a time-interval between the two related events: many non-teleological laws are about what happens during a period of time taken as a whole, e.g. the Law of Least Action. The difference beteween the two types of law seems to consist simply in the way in which the related variancy is discovered.

Here a comparison may be made with another type of law of a somewhat peculiar nature which occurs in psychology and in biology, and which shares with the teleological type the two characteristics that there is an interval of time between the determining and the determined event, and that the law holds under a wide variety of conditions which have been discovered inductively and not deductively. I refer to the laws governing what Bertrand Russell called "mnemic phenomena," laws which he called "mnemic laws."[15] The simplest example of such a law is that of memory recall, in which a present memory-image is determined (partially determined) by the occurrence in the rememberer of an experience of which the present memory-image is an image; but there are plenty of non-psychological examples in biology. The Mendelian laws of heredity state that sometimes some of the present characeristics of an organism are determined very precisely by the characteristics of its parent or parents at the time when the reproduction process commenced. More frequently the Mendelian determination is only statistical. In all the mnemic laws an earlier event is said to determine a later event without the intervening causal chain being specified or indeed known. We may postulate, if we wish, persistent genes to explain the facts of heredity, and traces in the brain or unconscious ideas in the mind to explain memory; but these are extra explanatory hypotheses going beyond what the mnemic law itself states. Or we may follow Russell's suggestion of supposing that there is a type of causation (which he called "ultimate mnemic causation") in which a past event directly determines a future event without there being any intermediate causal chain.[16] To suppose this would be almost as alien to our usual ways of thinking as to suppose that the future goal directly determines the present goal-directed action; and I agree with Russell in being unprepared to accept it if any way of escape is possible. As it is, physiologists are in the process of discovering strong-independent evidence for the existence of the genes which the geneticists postulate; and the neurologists or the experimental and clinical psychologists may in time discover satisfactory independent evidence for cerebral traces or for a persistent Unconscious. But in the meantime we have our mnemic laws; and the best account of them seems to me to be given by treating them in exactly the same way as teleological laws and making the inductively inferred variancy the distinguishing feature. This variancy is frequently large in the case of mnemic laws as it is in teleological: Lashley's rats retained their acquired skill in running his maze after large, and different, portions of their brains had been removed.

Both teleological and mnemic laws, then, assert that there is a causal chain connecting the determining and the determined events which holds under a wide range of conditions, i.e. that the system in question has a large variancy; and this variancy or plasticity has not been deduced from non-teleological or non-mnemic laws but has been established inductively by observation. The difference between them, that in a teleological law the determining event succeeds the determined event whereas in a mnemic law it precedes it, seems unimportant in comparison with their similarity; and I shall therefore class both types together under what I will for the lack of a better name call "biotic laws." I choose this name because the biotic laws which have struck our attention are those which apply to living systems, but my definition of "biotic law" includes no reference to life. Sometimes both a mnemic and a teleological law can be subsumed under one more general biotic law which is itself both mnemic and teleological: Mendel's laws of heredity state that characteristics of a set of organisms both are statistically determined by those of its set of parents and also statistically determine those of its set of offspring.[17]

This general notion of biotic law, of which teleological law is a species, is therefore offered as an attempt to settle the dispute between the biological "mechanists" and the biological "teleologists." But I fear that it will satisfy neither party. The teleologist will say that the whole account of teleological law in terms of causal chains and variancy of conditions presupposes the mechanist assumption that every event is physico-chemically determined, and that to admit teleological explanations *faute de mieux* is to ignore the essentially irreducible character of teleological law for which he is contending. The mechanist will declare that he has no use for teleological laws unless they are ultimate and irreducible; and that it is methodologically vicious to introduce new types of law just because we do not know all the laws of nature of the ordinary type. And both parties will join forces in criticizing my treatment as being unduly epistemological: the controversy, they will both say, is not as to how we derive our knowledge of general propositions about goal-directed activity, but is about the content of these general propositions; it is a question of the ultimate elements in the biological facts, not of the organization of our present biological knowledge.

All these criticisms, and the joint one particularly, are based upon what must be regarded as a naïve attitude to the function of a scientific law. For this function is just exactly that of organizing our empircal knowledge so as to give both intellectual satisfaction and power to predict the unknown. The nature of scientific laws cannot be treated independently of their function within a deductive system. The world is not made up of empirical facts with the addition of the laws of nature: what we call the laws of nature are conceptual devices by which we organize our empirical knowledge and predict the future. From this point of view any general hypothesis whose consequences are confirmed by experience is a valuable intellectual device; and the profitable use of such a hypothesis does not presuppose that it will not at some future time be subsumed under some more general hypothesis in a more widely applicable deductive system, nor that the facts which it explains will not some time be explicable by a quite different hypothesis in another deductive system.

Biotic hypotheses behave exactly like other scientific hypotheses in that they can frequently be treated as lower-level hypotheses in a new deductive system in which they are deducible from a set of higher-level hypotheses. For example, the special teleological law about a particular food (e.g. grass) as goal and a particular species of animal (e.g. horses) is deducible from a less special teleological hypothesis about food-seeking in general together with biochemical hypotheses about the conditions for the digestibility of grass. Frequently one goal-directed activity in an animal (e.g. the building of a nest) is followed by another type (e.g. sitting on the eggs laid in the nest); and the succession of these two types of teleological activity falls under some general teleological law (e.g. the mode of propagation of the species). Discussions as to the proper classification of instincts are largely discussions as to which is the best general deductive system containing higher-level biotic hypotheses for explaining the special instinctive modes of behaviour. When E. S. Russell puts forward the generalization that "the goal of a directive action or series of actions is normally related to one or other of the main biological ends of maintenance, development and reproduction,"[18] he is suggesting that a deductive system whose highest-level hypotheses include teleological hypotheses about these three ends will be able to absorb all the systems which are in terms of particular goals. Of course none of these more elaborate deductive systems have been worked out in detail; but the possibility of constructing them makes teleological explanatory hypotheses like the non-teleological ones in another respect also, namely, that we can hope to provide further explanations, and a deeper intellectual satisfaction, by incorporating special laws in a unified system.

I have given as part of the definition of a biotic law that it is not incorporated in a physico-chemical deductive system.[19] But if such a law, already incorporated in a biotic deductive system (i.e. one with biotic laws among its highest-level hypotheses), were to be found capable of physico-chemical explanation, we should not by that mere fact be estopped from continuing to make use of its place in the biotic system whenever we found it profitable to think in this way. The chemical deductive system with Dalton's laws of atomic combination as highest-level hypotheses has in recent years been included more and more within the more general deductive system of physics; but chemists find it far more convenient to treat most of their problems in terms of atoms and molecules than in terms of electrons or wave-functions. And there is one feature of both teleological and mnemic explanations which will almost certainly make them continue to be useful (whatever they would then be called) even if they could be superseded by physico-chemical ones. This feature is that usually teleological explanations make no reference to the exact length of time taken in attaining the goal, and mnemic explanations no reference to the exact length of time since the determining event. Indeed the unimportance of the time taken in reaching the goal is implicit in the persistency feature of goal-directed activity emphasized by biologists.[20] Teleological explanations do not specify the length of the causal chain, only that it attains the goal. So even if the biological mechanists can provide us with a complete explanation of life in physico-chemical terms, we shall probably continue to give teleological (or what would previously have been called teleological) explanations whenever the exact time taken in reaching the goal does not interest us.

I will conclude this chapter by summarizing the biological part of the argument. An account has been given of the distinguishing feature of teleological explanations which does not assume that such explanations are ultimately irreducible to chemistry and physics and which does not require any novel concept of causal law. To do this I have followed biologists in emphasizing the plasticity of goal-directed behaviour, and have analysed the peculiarity of a teleological explanation in terms of the related notions of the multiplicity of the causal chains by which the goal may be attained and of the variety of conditions under which the goal-directed activity may occur. These notions have been found to be also involved in mnemic explanation. What has been drawn is only an outline sketch which will need much working upon to make a convincing picture. But I have done enough to convince myself that what I have been trying to do is possible; and that the realm of biology will not have to sacrifice the autonomy proper to it if physics should succeed in establishing the claim, made by many biologists on its behalf, to be the Emperor of all the Natural Sciences.

[1]By saying that an event having the property B is *nomically determined* by an event having the property A, in conjunction with events having property A_1, having property A_2, etc., no more is meant than that the generalization "Every conjunction of an event having A with events having A_1, A_2, etc., is associated with an event having B" is a true generalization. Use of the language of nomic determination does not, therefore, presuppose any non-Humean analysis either of nomic or of causal statements.

[2]In the third ballet scene of *Le Malade Imaginaire*.

[3]The remainder of this chapter follows, with some alterations and additions, the text of my 1946 Presidential Address to the Aristotelian Society *(Proceedings of the Aristotelian Society, n.s., vol. 47 (1946-7), pp. i ff.)*.

[4]*The Directiveness of Organic Activities* (Cambridge, 1945).

[5]Jonathan Cohen has pointed out *(Proceedings of the Aristotelian Society, n.s., vol. 51 (1950-1), pp. 262ff.)* that such an explanation would differ from an ordinary causal explanation in that it is more difficult to specify the total cause of which the intention is only a part than it is to specify the total cause in an ordinary causal explanation. But this difference is only one of degree (as Cohen seems prepared to admit; loc. cit. p. 268 n.); and, however partial the intention may be as a factor in a total cause, it will not be later in time than the action which it is put forward to explain.

[6]W. McDougall, *An Outline of Psychology* (London, 1923), pp. 47f.

[7]Analyses of teleological explanations which have more or less resemblance to my analysis have been given by E. Rignano, *Mind*, n.s. vol. 40(1931), p. 337; A. Rosenblueth, N. Wiener and J. Bigelow, *Philosophy of Science*, vol. 10 (1943), p. 24 ("Teleological behvior becomes synonymous with behavior controlled by negative feed-back"); L. von Bertalanffy, *British Journal for the Philosophy of Science*, vol. 1 (1950), p. 157. The field of study called by Wiener "cybernetics" is largely concerned with "teleological mechanisms."

[8]*The Analysis of Mind*, p. 65.

[9]*The Directiveness of Organic Activities*, p. 144.

[10]K. S. Lashley, *Brain Mechanisms and Intelligence* (Chicago, 1929).

[11]The phrase "at all likely to occur" can be interpreted in different ways; but their difference does not affect the argument.

[12]In the case of a machine or of a laboratory demonstration a teleological explanation can of course be given of the action of a man in starting and controlling the machine or demonstration. Derivatively we can apply such a teleological explanation (as a "transferred epithet") to the working of the machine itself. But these explanations are all in terms of intentions as efficient causes, and so do not raise the special problem with which we are here concerned.

NOTES

[13]Jonathan Cohen, *Proceedings of the Aristotelian Society,* n.s., vol. 51 (1950-1), pp. 270, 292. Cohen holds that "a functional explanation asserts the explanandum to be [a] necessary condition (logically, causally or in any other generally recognised way) of the explanans and thereby also of the persistence under varying circumstances of a whole of which both explanans and the explanandum are parts" (loc. cit. p. 292). But the beating of my heart is not a *necessary* condition for the circulation of my blood: it is only because my anatomy includes a heart but no other mechanism for circulating my blood that it is causally necessary that the heart should *beat* in order that the blood should circulate.

[14]The term "functional explanation" may sometimes also be used to cover a mere description of the *modus operandi* of an organ like the heart. This would correspond to J. H. Woodger's third sense of "function" 7*Biological Principles* (London, 1929), p. 327).

[15]*The Analysis of Mind,* pp. 77ff.

[16]Ultimate mnemic causation would satisfy criterion II, but no stronger condition.

[17]Rignano's inclusion of the "finalist manifestations of life" in a "mnemonic property" goes far beyond my simple comparison.

[18]*The Directiveness of Organic Activities,* p. 80.

[19]By this phrase is meant a deductive system whose highest-level hypotheses are physical or chemical.

[20]For example, E. S. Russell, *The Directiveness of Organic Activities,* p. 110; "If the goal is not reached, action usually persists."

Ludwig von Bertalanffy

The following selection is taken from Ludwig von Bertalanffy's *Modern Theories of Development*, (1933), pp. 1-27, and reprinted from the 1962 edition with the kind permission of Harper and Brothers, New York.

MODERN THEORIES OF DEVELOPMENT

I
Biological Methodology

1. The Crisis in Biology

In the natural science of the present day we are witnessing a strange and disturbing spectacle. It is as though the grand sweep of its historical development, stretching from its beginnings in early Greek times up to the turn of the twentieth century, had to-day received a check. The foundations of our thought and investigation, hitherto regarded as assured, have collapsed. In their place new ways of thought, often paradoxical and apparently contradictory to the plain man, have appeared in bewildering variety, and among these still hotly contested ideas it is not yet possible to discover those which are destined to win an enduring place in our view of the world. Some years ago this state of affairs could be regarded as the break-down of Western science. But the remarkable developments which have recently been coming to fruition in physics suggest a totally different interpretation: we can see in the present state the raw and as yet unsettled early phase of a new step in scientific thought—the fruitful chaos out of which a new cosmos, a new system of thought will develop, albeit a view which will differ in essential points from that which we owe to Galileo, Kepler, and Newton.

In this place we need not describe the powerful revolutions which have occurred in mathematics and logic through the non-Euclidian geometries and the theory of aggregates, in physics through the Relativity and Quantum theories, and in psychology through the *Gestalttheorie*. The mere mention of these transformations suffices to indicate the place in the whole contemporary picture of the critical condition which we also find in the biology of the present day. When we speak of a crisis in biology it will be understood that we are not in any way saying anything prejudicial to its value. These general transformations in modern science signify rather the most powerful forward development which it has experienced since its foundation at the Renaissance. But it is at the same time essential that this state of affairs should be clearly reviewed, and that no attempt should be made to conceal it by entrenching ourselves behind theories which are now no longer tenable, or by shutting our eyes to the difficulties of our science.

> Modern biology is not in a position to display the results of systematic research in a system of concepts, or to represent the orderly behaviour which is common to its objects in a general theory. The place of theoretical science is taken rather by a heterogeneous multitude of facts, problems, views and interpretations. . . . Such a state of affairs cannot be improved upon by the piling up of new facts and opinions upon the older ones, but only by a fundamental re-organization after a process of careful sifting of those we already possess.

These assertions of Schaxel (1922, pp. 1 and 298) admirably express the present position of biology and its primary task. We find in biology a bitter dispute between spheres of investigation, opinions, and principles. In their methods and fundamental concepts the various branches of biology are extraordinarily diverse and disconnected, and occasionally even in direct oppositon to one another. The physico-chemical investigation of the vital process has given us, from the time of Harvey's fundamental discovery up to the most modern results of colloid-, ion-, and enzyme-chemistry, an uninterrupted chain of important discoveries—and yet there are good grounds for the belief that they still scarcely touch the essential problems of biology. The physiology of development and of behaviour work with systems of ideas which, at least at present, show only superficial relations to physics and chemistry. In genetics we have the most developed branch of biology, the only region in which we have an insight into the real biological laws, but we are still far from possessing a satisfactory theory of phylogenetic development, the fundamental idea of which is the most comprehensive that has so far appeared in the biological sphere. Attempts to master biology philosophically and theoretically are common enough outside the science, and stand in emphatic contradiction to its mechanistic point of view.

Whilst the majority of investigators find only physical and chemical processes in the object of their study, others find problematic metaphysical entities at the bottom of the vital phenomena. Between physicochemistry and metaphysics biology pursues a strange and crooked path. Because there is no generally adopted theory of the organism, a thousand different individual opinions, personally coloured in varying degrees, confront one another, among which a given worker will choose according to his personal taste and the requirements of his special sphere.

It is not our intention to describe in detail in this place the numerous controversies underlying the great biological theories of the last century, such as Mechanism, Vitalism, Selection Theory, Lamarckism, and Theory of Descent.[1] Under the influence of these theories, doctrines once belonging to the "assured acquisitions" of biology were established but have since been as much shaken as the seemingly "matter of course" ideas of space and time, of mass and causality, in physics. The above remarks will perhaps suffice to justify us in some measure in speaking of a state of crisis in biology.

But how can we speak of a crisis in this science when our knowledge of vital processes is being increased every year by a multitude of publications? It might be said that all such general conceptions are more or less fragile: let them go. We need not waste regrets over philosophical or semi-philosophical constructions. True science consists only in the knowledge of facts, and even the bitterest opponent of science cannot deny that this grows daily or even hourly.

Many investigators will perhaps adopt this attitude towards the state of uncertainty regarding fundamental doctrines in biology to which we have alluded. The empirical investigator is apt to look down upon "theory" with more or less disrespect, and therefore may not feel much distress at the uncertainty of the great theories.

But the empiricist is apt to forget two things. He forgets, in the first place, that a collection of facts, be it never so large, no more makes a science than a heap of bricks makes a house. In his scathing *Schöp-*

fungsliedern Heine makes God say: "Allein der Plan, die Uberlegung, da zeigt sich's, wer ein Meister ist." Only if the multiplicity of facts is ordered, brought into a system, subordinated to great laws and principles, only then does the heap of data become a science. Secondly, he forgets that no empirical science is even possible save on a basis of theoretical assumptions. Schaxel remarks very appropriately that "The empiricist moves hesitatingly between different attitudes. He wants to seem free, and yet is dependent upon ideas adopted at second hand with insufficient understanding." (1922, p. 5.) Thus the procedure of the biology of yesterday has failed: on the one hand "theory" has been looked down upon, and on the other, fact and theory have frequently been confused in an arbitrary and subjective manner.

A resolution of the present critical state of biology can thus only be sought in a theoretical clarification. Theoretical thinking must be recognized as a necessary ingredient of science. In biology until to-day such recognition has been rare, but in physics—which is taken as its model—it has always been a generally adopted demand. So much for criticism. Our critique will consist rather of construction, since we shall try to show a way to a new organization of biology which, we believe, will permit the present difficulties and contradictions—or at least many of them—to be overcome.

2. The Tasks of Theoretical Biology

If biology is to emerge from the crisis of its foundations and the accumulation of unrelated facts, as a critically purified exact science, the attainment of an assured theoretical biology will be necessary. But the term "theoretical biology" has two meanings denoting two different, but not completely separable, spheres of knowledge.

Theoretical biology in the *first sense* is the logic and methodology of the science of organisms. It establishes the foundations of biological knowledge and thus forms a branch of general logic and epistemology, whilst it may also be important for biological investigation. Problems requiring logical investigation, e.g. that of teleology, of the relation between fact and theory, of the significance of experiment in biology, &c., may be of the greatest importance for the whole direction of research in biology. Critical methodological clarification may constitute an active protection against the fallacies of hurried hypotheses.

But theoretical biology in the *second sense* signifies a branch of natural science which is related to descriptive and experimental biology in just the same way in which theoretical physics is related to experimental physics. That is the task of a theory of the various single branches of the vital phenomena, of development, metabolism, behaviour, reproduction, inheritance, and so on, and, in the last resort, of a "theory of life," in just the same sense in which there is a "theory of heat," a "theory of light," &c.

Since what has hitherto been called "theoretical biology" has consisted in great part of philosophical speculation, and since theoretical biology in the "first sense" consists of logical investigations, something must be said in clarification of the relations between theoretical biology and philosophy. As we have already mentioned, theoretical biology ("second sense") is just as much a branch of natural science as theoretical physics, i.e. it deals exclusively with the exact theoretical systematization of facts, and has no place for speculations. This point requires emphasis because voices are often raised in biology in rejection of theoretical biology as "merely philosophical" or "speculative" and superfluous. Such objections are entirely justified against many "theoretical biologies," especially those of a vitalistic character, which, however, are to a great extent "philosophical" and speculative and do not constitue scientifically applicable theorizing. But such objections are totally unjustified against theoretical biology conceived as a legitimate branch of natural science in the manner described above.

Naturally, it is not suggested that theoretical biology in the first and second senses, logic of biology and theory of life, should be regarded as totally unrelated to one another. Such a view would rather misrepresent the nature of theoretical science. Just as it is scarcely possible, in relation to the fundamental questions concerning space and time, action, deterministic or statistical law, &c., to draw a sharp line between physical theory and theory of knowledge, so will it also be the case in biology, in which the most general concepts (first of all that of "organism") on the one hand require logical clarification, and, on the other, form the foundation of biological explanations and theories. Such general scientific assumptions must be clarified in close connexion both with logical and epistemological considerations and with the empirical study of the relevant phenomena. It need hardly be mentioned that, like the fundamental questions of physics, those of biology, such as Vitalism and Evolution, touch upon philosophical and cosmological problems of the most important kind.

If we are to overcome the state of crisis in biology which we have discussed above, we require theoretical biology in both the "first" and in the "second" senses. We must first of all make clear to ourselves the methodological principles which must be applied in the different branches of the system of biological sciences. In doing this we shall be carrying out the task of theoretical biology in the "first sense" (Chapter I, 3-4). Then we must endeavour to reach a sound basis for a theory of life (Chapter II); and finally (in the main part of this book) we shall try to carry through the proposed programme of theoretical biology in a particularly suitable example, the phenomena of development. We shall endeavour to sift the current theories in this sphere and bring into application the theory we have traced in the general considerations.

3. The System of Biology

The attempt to arrange the various spheres of biology in a general system can be carried out in the following way.[2] We distinguish three stages in the system of biology.

1. Every science begins with an exact description and classification of its objects. Hence at the beginning of biology stands *systematics,* the aim of which is to give a catalogue, as complete and exact as possible, of all kinds of animals and plants. Related to this is the exact description of the different living forms or *anatomy* (including microscopical anatomy). *Comparative anatomy* and *morphology* result from the comparison of the structure of different organisms. Finally, in addition to classification in a system, in addition to simple and comparative description of living forms, the description of their distribution in space and time is necessary. In this way we have *bio-geography* and *palaeontology.* These two sciences are—to use Meyer's expression—not logically pure, but logically complex, since they involve oecological and phylogenetic problems, in addition to simple description of distribution in space and time.

2 *a.* After the objects of biology have thus been described and classified there remains the demand for a description of organic *processes.* It is clear that every vital process must first be *causally* described, and, if possible, by the method of causal explanation employed in the more advanced sciences of physics and chemistry. This is the method of investigation followed in *physiology.* About the conceptual methods of the physico-chemical investigation of life little need be said. It is clear that "the methods of the physiological chemist are peculiar

only in very few cases. They are almost exclusively taken from the neighbouring sciences of chemistry and physics" (Abderhalden). It is also widely believed that since biology in general coincides with the physical and chemical investigations of vital processes there is no necessity for peculiarly biological points of view, or for a special theoretical biology.

2 *b*. We believe that this view is not correct, since there are vital phenomena for the description of which other points of view are required. The first of these special biological points of view is the *organismic*.[3] We can undoubtedly describe the organism and its processes physico-chemically *in principle,* although we may still be far removed from reaching such a goal. But as *vital* processes they are not characterized in this way at all, since what is essential in the organism—as will be shown later—is that the particular physico-chemical processes are organized in it in quite a peculiar manner. We need not delay by entering into details in this place, and the reader may be referred to the discussions of Ungerer (1919, 1922, 1930), Rignano (1926, 1930-1), Sapper (1928), and Bertalanffy (1929). Whether we consider nutrition, voluntary and instinctive behaviour, development, the harmonious functioning of the organism under normal conditions, or its regulative functioning in cases of disturbances of the normal, we find that practically all vital processes are so organized that they are directed to the maintenance, production, or restoration of the wholeness of the organism. On that account the physico-chemical description of the vital processes does not exhaust them. They must also be considered from the standpoint of their significance for the maintenance of the organism. And we see that in fact—in spite of the postulate that science must only proceed physico-chemically—biology has at all times applied organismic ideas, and must apply them, and that whole spheres of investigation are concerned with the establishment of the significance of the organs and of organic processes for the whole.

The notion of "organ," of visual, auditory, or sexual organ, already involves the notion that this is a "tool" for something. As soon as we say that an animal has legs "in order to" run, the giraffe a long neck "because" it browses on the leaves—modes of expression which cannot be avoided in biology—we have already introduced a point of view which characterizes the significance of the organ for the maintenance of the organism—an organismic point of view. This point of view cannot be avoided so long as we cannot exclude the notion of an organ as "serving" some definite purpose. Similarly, the concept of "function" has an organismic sense: it only has significance within an organism, to the maintenance of which the function is exerted. We thus find *physiologic-al anatomy* to be the first branch of biology which investigates the organs in connexion with their functions, in their so-called "purposeful-ness" for the maintenance of the organism. Physiological anatomy fur-nishes a continual demonstration of the necessity of an organismic point of view in biology. As a second such branch we have *oecology*, which investigates the organic forms and functions as adaptations to their inorganic and organic environment. But since such concepts as disease, norm, distrubance, &c., are only significant in reference to the mainte-nance of an organism, *pathology* also belongs to the sphere of organism-ic branches of biology, but it is a logically complex discipline, since simple description and physiology have an important place in it.

For us there is no doubt that an organismic point of view of this kind is unavoidable. Organisms, as Kant knew, force this point of view upon us. It provides "a means of describing the organism and the vital processes from an aspect which is not touched by the causal standpoint" (Ungerer, 1919, p. 250). Indeed it might be said that the real biological problem lies just in this question of the significance of organs and vital processes for the organism. The best proof of the necessity of organicism and the insufficiency of the purely causal point of view is that mechanism also, contrary to its express declaration that only the physico-chemical causal standpoint is scientific, nevertheless cannot escape the use of "teleolog-ical" notions . . . Thus the mechanist Plate, in reply to the objection that "the purposefulness of the organic is not a problem for research" and that "exact investigation is only concerned with the search for causes," rightly says:

> The attempt to disavow the purposiveness of the organic as a problem for investigation leads to an arbitrary restriction of biology; for the latter must investigate and explain *all* relations of organisms, and hence one of its chief tasks must be to analyse and explain causally the great difference which exists between living and non-living natural objects. (1914, p. 31).

In modern biology there is, however, a strong movement in favour of excluding the "teleological" point of view as unscientific. In the first place the occurrence of dysteleology is brought forward as an objection. It is pointed out that even in organic nature by no means everything is "purposeful" or teleological. From the dystelelogical occurrences it is concluded that teleology only represents a subjective and an-thropomorphic point of view and that, in consequence, the physicochemical causal procedure is the only legitimate one in biology as well as in physics. This is the attitude of such authors as Goebel, Rabaud, B. Fischer, Needham, &c., who declare war upon the teleological point of view, whether it be Darwinistic, vitalistic, or purely methodological, and seek, or believe themselves to have already found, an ateleological standpoint.

Now, the refutation of this ateleological position has already been given in our foregoing discussions: we see that such a view would uproot whole branches of investigation, such as physiological anatomy, oecology, and pathology. We shall not here enter upon a detailed discussion of the problem of dysteleology (cf. 1928, pp. 83ff.; 1929). In any case it may be asserted that the attempt to refute the general "teleology" of the organic realm by picking out a dysteleological organ or process here and there is to pursue an "ostrich policy." There is a whole series of considerations by means of which dysteleology may be reconciled with teleology in general. In the first place the apparent uselessness or purposelessness of organic structures or processes may simply rest on the fact that so far no one has succeeded in discovering their "purpose," as was the case with the ductless glands before the discovery of internal secretions. Further, we cannot expect—even if we go so far as to assume that a purposive principle is active in the organism—that this is omnipotent. Even man, with his certainly "purposive" behaviour, is only able to guide matter for his use within certain limits. If, further, the organism is helpless before certain injuries—e.g. a minimal dose of prussic acid— this by no means represents a contradiction of organic teleology; for it is obvious that every system—including the organism—is only capable of existence in a definite environment. The possibility of an injury through unnatural interference no more refutes the "maintenance as a whole" of the organism than the fact that it cannot be filled with sulphuric acid destroys the "purposiveness" of a steam-engine. Moreover, ateleological reactions almost always occur—e.g. in the tropistic movements of animals—under experimental conditions which seldom or never occur in nature. That the teleology of the reaction is frustrated by the artifice of the experimenter is no more to be laid to nature's door as a defect, than intellect is to be denied to man because in a particularly difficult situation he does not choose to his best advantage. Finally, attention should be given in cases of dysteleology to the feature which Heidenhain has called "Encapsis": a process (e.g. suppuration of the brain) may be quite "purposeful" for a subordinate system of the organism, and yet destroy the system to which it is subordinate (as in the example mentioned, in which the skull prevents the escape of the pus), or we may have a situation in which the reverse is the case. At all events the most convinced representative of an ateleological point of view must admit that actually an enormous preponderance of vital processes and mechanisms have a whole-maintaining character; were this not so the organism could not exist at all. But if this is so, then the establishment of the significance of the processes for the life of the organism is a necessary branch of investigation.

The second main objection against the organismic standpoint goes deeper than the one just mentioned. It is said that only the causal point of view is strictly scientific, whilst "teleology" always involves the introduction of a mode of thought which is anthropomorphic and contradictory to the principles of science. Every "purpose" presupposes a striving, willing being, and to regard the mechanisms and processes in the organism teleologically means to assume a mystical anthropomorphic vital principle.

We can, however, say that the modern study of biological knowledge has succeeded in giving to the organismic point of view a formulation which avoids these objections. For this clarification we have chiefly to thank E. Ungerer (1919, 1922, 1930) who replaces the biological "consideration of purposes" with that of "consideration of wholeness." Applying the organismic method means in this sense investigating the vital processes with a view to discovering how far they contribute to the maintenance of organic wholeness. Ungerer points out that the so-called "purposefulness" of organisms is a pure fiction; it is "as if" a "purpose" were followed in organic processes, namely the maintenance of the organism in function and form. This means nothing more than: it is "as if" this preservation was willed or intended; but in the "as if" there lies also the implication that nothing is or can be known of the "willing" and "intending" nor of a willing or intending subject. Since only the maintenance, production, and restoration of the organism as one whole in function and form appear as "purpose" in the organic, or the special relation of a partial function to the total function of the whole, the "consideration of purpose" is to be replaced by that of "wholeness."

The teleological point of view in the sense here intended is quite free from hypothesis. It cannot be sufficiently emphasized that nothing mental is presupposed nor is a law of the purposiveness of reactions put forward, nor is it even asserted that the vital process must in all cases proceed in such a way as to attain the highest degree of purposiveness. . . . The confusion of the notion of wholeness with vitalism, which has done so much to prevent the attainment of a clear grasp of the present problem, must be fatal and lead to a disregard of the progress made so far, when such confusion is found in the works of one of the most important defenders of vitalism (Driesch!). It is therefore important to work out clearly the non-hypothetical, purely descriptive concept of wholeness and to show that it is the kernel of all "teleological" concepts of botany. The prejudice which sees in all teleology a concession to vitalism, the one-sidedness which overlooks an essential and fundamental feature of all living things and presents the shield of Darwinism to every true "purposiveness," must be just as strongly opposed as the unjustified attempt to seek a *deus ex machina* behind all "phenomena of adaptation." Quite apart from all controversies about mechanism it must be shown that the facts relating to organisms cannot be represented with purely causal concepts alone, and never were so represented, since a scientific elaboration of facts has always involved and still involves the use of the notion of wholeness. It remains to show what the "teleological point of view" means when freed from all admixtures, and how its use is unavoidable and free from danger." (1919, p. 39, f. iv.)

The strict mechanist Winterstein (1928) is of the same opinion.

When we attempt to conceive the vital occurrences of an organism we are at once confronted with the fact that we shall not succeed if attention is confined to the single processes going on in it at a given moment. We can only reach a satisfactory understanding if we consider them as partial processes in relation to the whole "vital mechanism." This inner connexion of the particular processes with the working of the whole gives the impression that the former are related to a total idea in a manner analogous to the way in which our own purposive behaviour appears to be guided by a preconceived plan. It is, however, possible to give a clear meaning, free from all subjective interpretation, to the originally anthropomorphic notion of purpose: "purposive" is nothing else than a short expression for all phenomena upon which the maintenance of an observed state or process depends.

Zimmermann (1928), Jordan (1929), Bertalanffy (1927, 1928) have reached views similar in principle to the above.

We thus see that organismic description of vital processes does not in any way constitute an "explanation"; it leaves the question open of how the maintenance of organic wholeness is achieved. The organismic standpoint neither asserts nor denies that the processes through which this is brought about are reducible to the phenomena of inorganic nature. Vitalism has erred in hypostasizing the concepts necessary for the teleological description of vital processes into active natural factors or entelechies.

The organismic point of view prejudges nothing regarding the theory of life, but every theory of life must of course give account of those features of organisms which this point of view reveals. On the other hand, the assertion that the organismic approach is incapable of leading to positive results is erroneous. In the first place, the teleological judgment of functioning organs or vital processes does not consist of popular wonderment—how beautiful and purposeful are all organic processes—but, like physics and chemistry, it promotes a thorough and, where possible, experimental study; on the other hand, as Winterstein rightly points out, the *a priori* assumption of the existence of regulations (thus the organismic point of view) has proved itself to be a principle of research of great heuristic worth: we may, for example, recall the fundamental ideas of Bier on the therapeutic value of fever, or Abderhalden's conception of defensive enzymes regulating the constancy of chemical composition.

In any case organismic description in the realm of the organic is just as necessary as the causal and physico-chemical; there is no sense in attempting to dispute away the organic character, the proper procedure is first to investigate, and secondly to explain it.

2 *c.* Alongside the causal and organismic there is yet a third form of description necessary in relation to organisms. This third form is the *historical*, which describes the organic forms and processes as products of an historical development. It is the business of *phylogeny* to provide such a description by establishing lines of descent. This historical point of view also represents a non-physical principle which forces itself upon us in the organic realm. On the other hand, in the sense here intended it only signifies a point of view and not a hypothesis. In order fully to understand organisms it is just as necessary to regard them as members in a process of historical development as it is to treat them as physico-chemical systems and as organic unities. And this general historical standpoint also is free from hypothesis, for, if we arrange organisms in phylo-genetic series and so regard the organic event as an historical process, we do not need for this purpose any assumption regarding the nature of life and its development.[4] Moreover, the historical standpoint serves to acquaint us with organisms from an aspect which is not touched by physics and chemistry; hypotheses only begin when we reflect upon the *causes* of development, upon the nature of historical accumulation, &c. . . .

From the comparison of the multiplicity of living forms and processes certain uniformities result which are the same in all, or in very many, organisms. The bringing together of these uniformities is the task of *general biology*. Or, we can say: general biology is the collection of general rules which can be derived from the consideration of the multiplicity of vital phenomena. Such rules issue both from the comparison of forms in morphology, and from the description of vital processes from the causal, organismic, and historical standpoints.

2 *d*. We have now indicated the necessary presuppositions for an adequate description of the vital processes. All these points of view—the physico-chemical, the organismic, and the historical—represent, as we have said, exclusively methodological assumptions. We remain in the domain of the establishment of empirical facts just as much when we study the vital phenomena physico-chemically, as when we investigate them organismically or historically. In neither case is anything hypothetical asserted about the nature of life when we apply these standpoints. The domain of hypothesis is theoretical biology, which is necessary on the one hand for the general explanation of great spheres of fact, and on the other for making a science of law possible.

But since the physico-chemical point of view does not suffice in biology, but must be supplemented from the organismic (physiological anatomy, oecology) and historical (phylogeny) standpoints, the necessity of biology as an independent science—at least as far as its descriptive stages are concerned—is proved. For even a complete physico-chemical description of the organic processes would not—as we have seen—render the organismic and historical points of view superfluous. They would always remain necessary avenues of approach for the description of vital phenomena.

These views are, however, in opposition to a widespread opinion which—supporting itself on an assertion of Kant[6]—only regards the causal and, in the last resort, mathematical study of phenomena as "scientific." Against this it must be urged that science must always relate itself to the facts, and is not in a position to prescribe to reality what can or must be the case. If organic "teleology" and history represent essential features of reality, then science must take account of them, and, in order to do this, it does not need the permission of a dogmatic epistemology. To forbid the investigation of certain features of reality is to set up a wholly inadmissible restriction of science. If the vital phenomena present features which do not lend themselves easily to theoretical treatment by the means which have been devised in other branches of natural science, the proper procedure for biology would seem to be to devise its own technique for dealing with them; not to ignore them or to restrict itself by arbitrary definitions based too naïvely and exclusively on traditional models. It is naturally impossible to "refute" a definition. Every one is at liberty to define "science" as he pleases. But the least requirement which can be expected of a definition is that it does not fly too much in the face of the actual state of affairs. But this is precisely what is done by a definition of science which makes it equivalent to mathematical physics. According to this definition not only are the "mental sciences"—psychology, sociology, history, &c.—not sciences, and never can be such, but the same will be true of large branches of natural science as well. If this programme is to be realized it will be necessary to displace the majority of professors of zoology and botany, the systematists, anatomists, morphologists, physiological anatomists, oecologists, and students of phylogeny, because they do not pursue physico-chemical, and hence "scientific," investigations at all. In any case, besides the causal and physico-chemical investigations of living things, morphology, oecology, and phylogeny represent legitimate branches of science, and the equating of "science" with mathematical physics seems in any case, in the light of the actual state of affairs, to be absurd. In this connexion we may quote the comments which a supporter of our view, the distinguished physicist and natural philosopher Bavink, has made on the views of the author given in the foregoing paragraph.

In my opinion they finally and irrefutably dispose of the fatal error, which has injured an epoch of scientific thinking, of equating science in general with mathematical physics. . . . Science is any attempt to bring facts into logical order. Mathametical physics is only one special aspect of this activity. That the mental sciences do not proceed in this way has long been clear. Now we see that such a narrow definition does not even suffice for natural science." (1929, p. 340)

3. But the task of scientific biology is not yet exhausted by the topics so far mentioned. In them it is only a question of the establishment of relations of facts, whether these are of causal, organismic, or historical nature. We remain at the "descriptive" level of science. The regularities among events established here find their expression in "rules" or "empirical" laws. Such empirical laws are unrelated among one another; they are not deducible from higher principles; we can state no necessity for the occurrence of just these regularities and no others. If we wish to bring such empirical laws into relation, if we wish to "explain" the particular occurrences and the rules they exemplify, we require hy-

pothetical ideas for this purpose. A strict system of law signifies a logical connexion of conceptual constructions. On that account they cannot—as we shall see in the next section—be simply read off from experience. As relations within a conceptual construction the natural laws are deducible from laws superior to them and admit of subordinate laws being deduced from them; as such they possess logical necessity if the premisses from which they are deducible are agreed to.

A brief consideration may be in place here concerning the question of the relation between "description" and "explanation," and here we may refer the reader to the admirable discussion of Bavink (1930, pp. 23 ff.) According to this author's definition, a hypothesis is "the supposition of a general state of affairs as underlying certain special phenomena occurring in experience, from the presence of which and its assumed laws the phenomena of the region of fact concerned can be deduced." If this is the case it must be the aim of science to establish hypotheses directly whence the original hypothesis becomes a proved fact, as was the case with atoms and light waves which thereby have come to have the same "reality value" as "stones and trees, plant-cells and fixed stars." With the aid of this definition we can express the relation between "description" and "explanation." If description is the simple assertion of facts, explanation signifies the logical subordination of the particular under the more general, the systematization of the given facts by means of general connexions. This also means that every explanation again demands a new explanation, i.e. the search for still more comprehensive connexions, in relation to which it appears as "description," as the establishment of a matter of fact.

The first task of theory is thus to give a common explanation for a series of otherwise unconnected facts. Secondly, the hypothetical ideas which theoretical science elaborates make possible the setting up of a system of strict natural laws. We see this double significance in the most fully extended theorems we possess, those of physics; for example, the electron theory makes possible an explanation of an extraordinary number of single phenomena, and on the other hand it has led to the establishment of laws for these phenomena.

Thus in addition to the realm of perceptions there is in theoretical science a second realm of hypothetical structures between which the relations of the natural laws hold. In what relation this "second realm" stands to the "third realm" of metaphysical reality is a question which the scientist need not answer, but may leave to the general theory of knowledge; modern physics, however, may be able to give some hints in this direction:

> In opposition to a widespread view it is without significance for physics whether we call the content of the first realm (sense-data), e.g. the perceived colour blue, mere phenomena, and that of the second, e.g. the corresponding electromagnetic vibrations, "reality" in the realistic sense, or whether, on the other hand, in the positivistic sense, we call the first the "really given," and the second as only consisting of conceptual complexes of those sense-data. On that account physics does not say: "where this blue appears there is, in reality, such and such an electronic process," nor "in the place of this blue we conceive such an electronic process in order to make calculation possible," but physics expresses itself quite neutrally with the help of purely formal co-ordinating relations, and leaves the question of further interpretation to a non-physical investigation. (Carnap, 1923).

In any case the theoretical constructions must be so constituted that they are, in Schlick's phrase, "unequivocally coordinated" with the perceptual world. If that is achieved, the fulfilment of the principal task of science—the exact prediction of future events—is made possible with the help of natural laws. If our conceptual constructions, the theoretical structure and laws, are unequivocally co-ordinated with the phenomena, it is then no wonder that they not only fit the past but also future events, and hence enable us to "prophesy" the future.

Scientific law does not consist, as is often said (Dubois-Reymond, Sigwart, Roux, and others), in an insight into the "causal necessity" of the events. The striking refutation of this definition is provided by modern physics, which recognizes the impossibility of a causal determination of the ultimate intra-atomic processes, and regards all natural law as purely statistical. It is not an insight into the *causal* necessity of the processes which gives physical laws their strict character, but rather the insight into the *logical* necessity of those laws; in fact, according to Bavink (1930, pp. 60 ff.), the causal relation is itself reducible to logical necessity.

It thus comes about that theoretical science is at the same time science of law, and only as such is science of law possible at all. In physics and chemistry this has long been attained, but in biology, on account of the widespread aversion for theoretical thinking, we are very far from such a state of affairs. Nevertheless, or just for that reason, we must also demand a *theoretical biology* as the crown of the whole structure of the science of life—it being presupposed that the biological happenings are not exhausted by the simple physico-chemical description of the individual processes into which, in a given case, it is analysable—an assumption which we can without difficulty prove to be incorrect . . . The chief task of theoretical biology will be to explain the general, organismic, and historical character of biological events from general assumptions. The great systems of mechanism and vitalism represent such theories of life, but we shall see that in their place a more satisfactory foundation must be sought for biological theory.

In this way we have reached a survey of the chief directions of biological investigation. We see that biological knowledge operates at three levels: in the first level it deals with the ordering, the simple and comparative description, of its objects. In the second the causal, organismic, and historical connexions of the organism are investigated, and—in "general biology"—rules are set up for the uniformities which here present themselves. The third state—that of theoretical biology—yields, with the help of hypothetical suppositions, the laws of biological processes.

It need not be emphasized that in the foregoing analysis we have not indicated branches of science which in practice are strictly separable, but rather various attitudes. Consequently a given piece of scientific work rarely belongs to only one of the spheres distinguished but usually embraces several of them. But these attitudes must be clearly distinguished from one another from the logical standpoint. If this is not done we have "romantic biology" (Schaxel), in which organismic descriptions masquerade as causal explanations, a supposed mechanism will appear in spite of continued use of non-mechanical teleological and historical notions, and theory and fact will be inextricably confused.

4. The Method of Theoretical Biology

To-day biology is still in its *pre-Copernican* period. We possess an enormous mass of facts, but we still have only a very incomplete insight into the laws governing them. Apart from genetics, which approaches most closely to the goal of theoretical science, the most superficial glance serves to show that whilst in physics we speak everywhere of "laws," in biology this is the case only in rare and isolated instances. The absence of laws rests on the fact that although we have had numerous biological theories we have so far had no theoretical biology. Theoretical science and science of law are one and the same, and the lack of theoretical biology prevents us from taking the step from a purely empirical descriptive science to one of exact laws.

We must therefore consider the important question: by what means is an assured theoretical biology to be reached? A further comparison with theoretical physics will again serve to throw into relief the peculiar state of theoretical biology. For whilst theoretical physics is a completely developed science, in fact the most highly developed one, theoretical biology has not emerged from its swaddling clothes. For this reason the latter is still faced with a task which for physics has long ceased to be of great importance. This first task of theoretical biology is the *critical analysis of the various theories* which have hitherto been put forward in relation to the various vital phenomena, in order to discover which may claim the title of exactitude. We already have such a multitude of hypotheses and theories—often in all thinkable logical shades—that it seems desirable to make a critical survey of these first before attempting to add new ones to them. To establish what we already possess by way of really firm general knowledge about organisms is the first task of theoretical biology, which we can call "analytical" theoretical biology.

But naturally the task of theoretical biology is by no means exhausted by such a sifting of theories. Its last aim is to establish a unitary *system*. And in order to be clear about what this entails, it is necessary to inquire into the way in which a scientific theory is built up. For this purpose we can make use of the excellent account given by Kraft (1926).

It is a common opinion that the principles underlying scientific theories are to be derived "from experience." But a closer study of highly developed theoretical systems shows the complete falsity of this view. The principles of mechanics, for example, cannot be empirical propositions because they involve relations which, in such a form, are never met with in experience. The fall of bodies in accordance with the law of gravitation is an ideal process. If we make cinematograph films of falling bodies and measure from them the time and space involved we could only find an approximate conformity to law, never an exact one. Scientific theory, as exemplified by mechanics or theoretical physics, has the character of a *hypothetico-deductive system*. Freely chosen ideal assumptions are first clearly stated, and then, by the introduction of special conditions, consequences are deduced with logical rigour from these, and such consequences are then compared with experience and, if the premises have been suitably chosen, are thereby verified. The rigorous character of the scientific theory rests *only* on this procedure.

The antithesis to theory is inductive science. But all inferences from experience, since they rest on particular facts showing only an approximate regularity, can never be more than "assumptions." All general knowledge about reality is rational construction, and theory and induction are distinguished only by the fact that the latter proceeds from the facts and the former is verified by them.

An insight into the essential nature of scientific theory is of great importance from the biological standpoint. We see how completely untenable, both logically and psychologically, is the view that natural laws can simply be read off as a result of recording as large a collection as possible of empirical data. This view is logically untenable because natural laws are not found running about wild in nature, but must be reached by a process in which abstraction is made from all "perturbations." Psychologically this is only possible by means of a happy intuition which is able, with the eye of genius, to discern the essential features of an event behind the complex multiplicity of phenomena and produces the hypothetical statement which brings the facts into order as it were at one stroke. This is true whatever view we may take in regard to the status of the "second realm" of theoretical science, i.e. whether we regard it as establishing real relations, or as merely a set of conceptual constructions in the positivistic sense. We do not wish to burden the present study with this question, but will refer the reader again to the excellent discussion given by Bavink (1930).

Newton saw the famous apple fall from the tree. Here, if anywhere, we can speak of an intuitive grasp of the general law in the particular case, of a "Wesensschau" in the sense of Husserl. But this primary intuition is necessary for the setting up of *every* law. Many apples fall from trees, but only rarely is there a Newton to apprehend the laws of the world in such events. Hundreds of thousands of apples, registered with every possible accuracy, would never yield the great law of gravitation. It is a foolish hope to suppose that by the accumulation of innumerable single cases great laws will finally emerge, like Venus from the nebulous sea.

The significance of all this for biology is obvious. It is not true that empirical knowledge, however extended, suffices for the foundation of a well-systematized science. The latter can only be reached by the close co-operation of experience and deductive-hypothetical thinking.

We are indebted to Kraft for another important notion. He rightly points out that mathematics (as ordinarily understood) is by no means the only possible foundation for a strict theory, i.e. a hypothetico-deductive system, but such is always present when we have deduction from idealized conditions, as is the case, e.g. in Menger's deductive economic theory. It is not impossible to suppose that in biology, in which deductive theory in the mathematical form is as yet scarcely possible, such a system would be appropriate in a non-mathematical form, leaving open the possibility of subsequently fitting such a theory into the system of mathematical logic to be extended in the future.

At the present day, however, the necessity of theoretical biology is by no means generally recognized. The view that experimental investigation alone can claim the title of "scientific" still reigns. For this reason a few words about the functions of theoretical biology in its two chief aspects will not be out of place.

The importance of its first task—the critical analysis of existing theories and concepts—should be sufficiently obvious. All the critical phenomena in biology which are brought to light in the course of this discussion—the intermingling of contradictory points of view and theories, the lack of a generally accepted theoretical system, the survival of theories which have long become obsolete, the muddles and contradictions involved in many biological concepts—all these can only be overcome by means of analytical theoretical biology.

Not less important is the second, constructive function of theoretical biology. From what has been said in this section, it follows—as Physics shows so clearly—that theoretical and experimental science are necessary complements of one another. A systematic science can only be constructed by their mutual co-operation. A science only becomes a science of exact laws when it becomes theoretical. The ideal of "science without hypothesis" is quite justified if it means the rejection of superfluous speculations, but it is a mere phantom if it intends to suggest that any science is possible without a framework of theoretical concepts. (Cf. also Schaxel, 1922, pp. 234, 298, &c.)

A majority of biologists at the present day profess to reject "theory." Nevertheless, while paying every respect to the importance of experimental investigation, we cannot wholly agree with the frequently repeated demand for more experimentation and less theorizing. When we open one of the biological reviews, and glance at the thousands of experimental investigations which are published yearly, we cannot avoid the heretical opinion that it is perhaps not so necessary to add another dozen or so to these as seriously to set about the task of theoretically exploiting the mountain of raw material we already possess. It may be objected that biology is not yet "ripe" for such an undertaking. On the contrary it is essential to understand that empirical investigation and theory can only grow properly in correlation with one another, and that the assumption that theory is only possible and necessary when the collection of data is finished is quite erroneous.

We must not conclude this defence of theory in biology without admitting that of course biologists have had many good grounds for their distrust of "theory." Nothing is so dangerous as the groundless speculation and theorizing in vogue among biological outsiders, but unfortunately this is not unkonwn even in the science itself. There is also another ground for the anti-theoretical attitude of contemporary biologists which is not difficult to understand and agree with. Only too often do we see the theorist leave the solid ground of experience and experiment and disappear into the blue mists of metaphysical speculation. When once the aversion to *this* kind of theoretical biology has seized biologists, it may easily happen that *every* kind of "theory" comes to be regarded as a departure from his proper scientific business. Here, then, is another point where there is a need for change in contemporary attitude, a change which ought not to consist in the rejection of theory in general but in taking seriously the need for a scientific theoretical biology, whilst at the same time declaring war upon all such light-minded speculation as has been responsible for the mistrust of "theory" in biology.

NOTES

[1] A review of these controversies is given in our *Kritische Theorie der Formbildung*, 1928.

[2] For other systems of biology see the discussions of Tschulok (1910), Meyer (1926), and Bertalanffy (1928, chap. ii).

[3] This word replaces the old term "teleological". It will be seen in what follows that "teleology" as we conceive it has nothing to do with any psychological or vitalistic assumptions which were often confused with this point of view.

[4] We say: the historical *point of view* is free from hypothesis. That many, probably most, of the ancestral series set up by its aid are extremely hypothetical is indeed obvious. Logically, we must, however, sharply distinguish two kinds of hypotheses. First, those which serve to bridge the gaps in our incomplete knowledge of facts; such are especially noticeable in phylogeny in consequence of the fragmentary nature of our fossil material, but they naturally occur also in the phyico-chemical and organismic procedures. Our ideas about the stages of assimilation of carbon dioxide, or about the significance of the Golgi apparatus in the cell, are still hypothetical in character. But the student of phylogeny hopes sooner or later to be able to demonstrate those members of the developmental series of man, for example, which are still lacking, just as the biochemist hopes to fill in the gaps in our knowledge of carbon assimilation, and the physiological cell-anatomist hopes to demonstrate visibly the secretion of the Golgi apparatus. But if the physicist introduces the "hypothesis" of material waves, or the biologist assumes that development occurs

through the mutation of genes, he does not attempt to fill in gaps in our knowledge of matter of fact, but hopes to explain the facts themselves. To use an expression to be introduced later [p. 00]: the "complemental" hypotheses express expectations about future experiences in the "first world" of sensible reality, the "explanatory" hypotheses belong to the "second world" of theoretical science. Such explanatory hypotheses are not involved in the mere setting up of ancestral series.

[5]For a criticsm of misunderstandings of Kant's assertion see the pertinent remarks of Woodger, 1929, p. 234.

J. H. Woodger

Joseph Henry Woodger is Emeritus Professor of Biology at the University of London. Born in 1894, he studied zoology at University College from which he was graduated in 1914. He taught at the University of London from 1919 to 1959 when he retired from teaching. Woodger has been Tarner Lecturer at Trinity College in Cambridge from 1949 to 1950. Some of his books are: *Biological Principles, The Axiomatic Method in Biology, Biology and Language,* and *Physics, Psychology, and Medicine.* He currently resides in Epsom Downs, and is working on an axiomatization of theories of genetics.

The following selection is taken from *The Quarterly Review of Biology,* Vol. V, No. 1, March 1930, pp. 1-22, and is reprinted here with the kind permission of The Williams and Wilkins Company, Baltimore.

THE "CONCEPT OF ORGANISM" AND THE RELATION BETWEEN EMBRYOLOGY AND GENETICS

Part I

It has become almost a commonplace of the times in which we live to demand that knowledge—scientifically garnered and sifted—should be applied in all human undertakings. It is no longer considered sufficient, as it was in former times, to be guided by "intuition" or rule of thumb tradition, however successful this may be. No. Nowadays it is held that the process in question, whatever it is, must first be subjected to scientific analysis, and the work carried out in the light of the knowledge thus obtained. Now it seems at first sight a curious fact that the very people who are responsible for this change, or at least for providing the knowledge upon which it is based, do not apply the principle to their own activities. That is to say, the people who pursue natural scientific knowledge may be said, as a rule, and from one point of view at least, not to know what they are doing in somewhat the same sense in which a cook may be said not to know what she is doing when she uses baking powder. She knows that with this ingredient the buns will "rise," and so long as this is the case why need she concern herself with the properties of CO_2 or the laws of gases? In the same way, so long as the investigator of nature continues to make discoveries (and the volume of papers issued in the quarterly journals shows no sign of decline) there would seem to be no occasion to deflect attention from the business of investigating nature to matters relating to the process of investigation itself.

But the situation is complicated in various ways which require disentangling. In the first place the biological student does in fact receive instruction in the use of instruments, dyes, etc., based upon knowledge obtained by a scientific study of such things, and he is also required to know a great deal about other branches of natural science which will be involved in his biological investigations. But there are other important points which are liable to be overlooked. It is generally believed that the business of the biologist is not exhausted when he has made his observations and recorded them in the quarterly journals. Those journals are so many store-houses of data in a raw state and, in a sense, in an "unavailable" form,—in a form, that is to say, in which they are of little interest or value to anyone but the small group of specialists in the particular branch of biology concerned. Before such knowledge can become of wider availability it must be subjected to a *further* process through which it becomes articulated into a wider body of knowledge and brought into relation with other branches of the subject. Thus there appear to be two different but related processes: investigation and interpretation, and these two processes appear to be different in their nature, their outcome, and in the "canons" which regulate them. The investigatory process reduces at bottom either to observing organisms or parts of organisms in their natural relations, or to altering their natural relations in a systematic way, and recording the results. And, since the primary aim of investigation is discovery, whatever procedure leads to discovery will be accounted *good,* and whatever methods fail to achieve this end will be rejected. Thus, as far as methods of investigation are concerned (whether this refers to technical processes or to intellectual tools, i.e., "working hypotheses") heuristic success will be the touchstone by which the investigator will measure all things.

So much, then, for the criteria by which we judge the value of a method of investigation. But how do we judge whether the *outcome* of such methods is "good"? The outcome of an investigation is a series of propositions which are conveyed to the world at large by means of printed sentences in the quarterly journals, and the primary requisite of such propositions is that they should be *true*. The majority of such propositions simply state what did in fact happen in a certain laboratory on a certain date under more or less definitely known conditions. We all know what is meant by saying that such records are true, and no honest man would include false propositions in his report if he knows they are false. But when we consider the process of interpretation and *its* outcome the situation is different. Here it is a question of intellectually mastering the data furnished by the former process—of systematizing the propositions already obtained in the form of general propositions which are hypothetical in the sense that we do not *know* that they are true, although we know that they are either true or false. And we are usually in the position that it may be possible to discover whether they are false, but it will never be possible to *know* that they are true. And by knowing I *mean* knowing, *sensu stricto*, as contrasted with supposing or believing.

Now it is clear that the sort of systematized knowledge about instruments and dyes, etc., which the investigator uses to guide him in his pursuits, is natural scientific knowledge. It is applied physics and chemistry. But the sort of knowledge required for the guidance of the process of interpretation will be knowledge about the properties of *knowledge itself,* and will not be natural scientific knowledge. It seems to be the case as a general rule that the people who pursue natural scientific knowledge never pay much attention to knowledge about knowledge itself, and the people who make knowledge itself an object of scientific investigation do not always know much about the subject matter of natural scientific knowledge. It is this state of affairs which is responsible for a situation comparable to that of the cook and the baking powder. Everyone recognizes the desirability of knowing something about the physical instruments used in scientific investigation, but the importance of understanding the properties of the intellectual tools involved—concepts, propositions, principles of inference, "working hypotheses," postulates, etc.,—is much less clearly appreciated. And from the standpoint of the process of *interpretation* this *may* be a misfortune.

An excellent illustration of the desirability of devoting some attention to these matters is furnished by the history of comparative psychology as related in a recent article in this REVIEW by C. J. Warden (11). The point of special interest in this example is that the position reached at the end of the story could have been grasped perfectly well at the beginning if more attention had been paid to an understanding of knowledge itself, and if certain speculative assumptions had not been permitted to influence empirical investigation. Such assumptions always have their "good" and "bad" sides. On the one hand they often lead people to make certain investigations which they would not otherwise have done, and hence lead to discoveries, and they are therefore important heuristically. But they are bad if they are allowed to influence the outcome of an investigation, if they lead to too great a restriction of the field of interest, and to a one-sided selection of evidence. All these points are illustrated in the paper referred to.

It is important to distinguish three kinds of assumptions. First, there are those commonly called "working hypotheses," and obviously the first requisite of such an assumption is that it *will* work, i.e. is such that it can be put to an empirical test which will definitely *disprove* it if it is false. We cannot put it more positively than that because, if the test does *not* disprove it, it does not follow that it is true. This happens, for example, in the case of hypotheses which assert the existence of entities which can never be observed. A good example of such a hypothesis which *could* be disproved is furnished by Weismann's hypothesis of development. But there are also hypotheses of a totally different character which are such that no experimental test can *possibly* be devised which would decisively disprove them if they were false. Of these there are two distinct kinds. First, there are those assumptions which, if false, would render nugatory all intellectual activity. In regard to these, although we have no *reason* for believing them to be true, we yet have a very strong *motive* for believing them to be so. In natural science we adopt certain assumptions of this kind: they are *postulates* necessary for the possibility of knowledge. The second type of untestable assumptions embraces those which are not necessary for the possibility of knowledge. These might be distinguished as *metaphysical* assumptions, and it is clear that if they are admitted into science their proper character should not be forgotten, especially if they are taken over uncritically from common-sense beliefs. Also regarding assumptions in general, it is evident that if you *assume* a certain proposition *p*, then any proposition which depends for its truth upon the truth of *p* must itself be hypothetical and cannot be regarded as one of the facts which are the outcome of your investigation. And if it is considered necessary to assume that *p* is true, if even the possibility of its falsehood cannot be entertained, then clearly the aim and outcome of your investigation cannot be to support or to deny *p*, but must be confined to propositions which will be true only if *p* is true. It is necessary to draw attention to these points because they are involved in what follows and have often been neglected.

There is another feature of natural science in regard to which "knowledge about knowledge itself" is important. We distinguish between experimental and theoretical physics but we make no such distinction in biology. There are biological theories in abundance but no theoretical biology. There are books in plenty on experimental embryology but none on theoretical embryology. Why is this? One reason seems to be that biological knowledge is believed to progress by "observation and experiment." It is commonly taught in popular books on the history of natural science that at the Renaissance someone discovered that observation and experiment were essential to the attainment of natural knowledge, and that it was this which distinguished the later centuries from the Middle Ages. This is perfectly true, but it is a half-truth, and half-truths are usually dangerous. What is here omitted is every bit as important and essential to scientific progress as what is asserted. If physics had proceeded on these lines there would never have been any theoretical physics, and if biology does not pay attention to what is here omitted it is doubtful whether there will ever be any theoretical biology. We therefore require to discover what this other ingredient may be. Theoretical physics is sometimes called *mathematical* physics, and this seems to have led some biologists to suppose that all we have to do is to "apply" mathematics to biological data and theoretical biology will automatically emerge. This is, I think, a mistake founded upon a too simple and too superficial view of the situation. It is not so much the mathematical laws of physics as such that we require, but the mathematical *method* in the broadest sense, and this is quite a different thing. Natural scientific knowledge springs from a fertilizing union of two "realms": the realm of sense-experience or perception, on the one hand, and the "logical realm" or the realm of abstract logical entities and relations, on the other. The mistake of the Middle Ages was to neglect the former and to attend to only a very restricted aspect of the latter; the mistake of biology has been to concentrate entirely on the former and to use the latter only unconsciously or "intuitively." The success of theoretical physics has depended on the fact that, aided by pure mathematics, it has been able to explore and combine both. But it is a great mistake to suppose that mathematics as ordinarily understood exhausts the logical realm. It is now recognized that it represents only certain aspects of it, and this fact has come to light in comparatively recent years as a result of investigations into the nature of pure mathematics itself. Knowledge about knowledge itself is very important for understanding the logical realm.

II

Biology at the present day seems to be passing through a critical period in which new ideas are clamoring for attention, whilst their recognition is being retarded by various conservative factors. In various quarters signs are discernible that this is slowly becoming realized. As far as *investigation* is concerned the turn of the tide took place some time ago, when what has been called the "romantic" period came to an end with the tardy recognition that experiment was required for further progress. But accompanying these changes in the standpoint of investigators there has been no fundamental change in biological *interpretation*. It is here that change is confronted with the most formidable obstacles. It is easy enough to appreciate the need for supplementing observation by experiment, but it is quite another thing to foresee the need for a revision of our ways of thinking, and a very difficult thing indeed to carry out such a revision. These things appear to be better understood in Germany than in the English-speaking countries. In England it is not among men of science that we find most interest taken in the newer biological ideas. But in 1919 J. Schaxel (10) published a book in which he pointed out the exceedingly heterogeneous nature of biological thought, drew attention to its inconsistencies, and urged the need for a critical sifting of its fundamental concepts. The first edition was out of print before the end of the year and a second was issued in 1922. It is difficult to imagine such a book passing through two editions in England. Schaxel's chief aim was to drive home the necessity of understanding that strictness of *thought*, as well as exactness of investigation, is essential if biology is to emerge from the morass into which its "careless *Begriffsromantik*" (as Schaxel calls it) has brought it. Since the first publication of this book Schaxel has edited a series of *Abhandlungen zur theoretischen Biologie,* to which a number of German thinkers have contributed, and it is among these that the reader may discern the signs of reawakening to which reference has been made. These writers all recognize that a given set of assumptions will commit you to a certain definite set of conclusions. But in biology our fundamental notions are for the most part so vague and lacking in precise definition that it is impossible to work out their logical consequences to any clear cut result, as any one will have realized who has attempted to discover the precise points at issue in the traditional biological controversies. The German authors make a distinction between "law" (Gesetz) and "rule" (Regel) which recognizes the distinction between what I have called the logical and empirical realms. By a *law* they mean an *a priori* logical system capable of strict deductive development, although this need not be "mathematical" in the narrow sense, but may be "logistic" in the sense of C. I. Lewis (9). By a *rule* these authors simply mean an empirical inductive generalization. But by no means all German biological writers either make this distinction in the above way or realise its importance. Thus B. Dürken (4) writes:

> It must never be forgotten that the task of the exact investigation of nature does not consist in trying to explain nature by deduction from general principles, and so adapting the facts to preconceived opinions; but lies in the attempt to obtain a general view of natural occurrences inductively from the greatest possible number of isolated facts.

Now this is certainly not the way in which physics developed, as illustrated, for example, by the procedure of Galileo and Kepler. It represents only one aspect of the process. Induction is necessary but not sufficient. It will give us biological theories but not theoretical biology. Only if and when a system of logical relations is discovered from which the empirical generalizations can be deductively developed and into which the biological concepts enter as values of the logical variables will anything approaching theoretical biology, which is at all comparable with theoretical physics, be possible.

The authors in Schaxel's *Abbandlungen* deal, for the most part, with the clarification of biological concepts, a most necessary preliminary, and also with the working out of new ones. And if it is possible to pick out one theme which is especially prominent in these discussions it seems to be one which turns on the question whether we are to regard the organism *as* an organism or not, and on the difficulties which present themselves when we answer this question in the affirmative. To an outsider it will seem strange that there should be any such question, but it admits of course of a simple historical explanation. In the past the concept of organism has not been employed by the majority of biologists, but instead, owing to the methodological success of the notions bequeathed to us by René Descartes, organisms have been more commonly regarded as, in some sense, machines, although in what sense has not always been clear. I have attempted to clarify this question elsewhere (17). To discover what the "concept of organism" means at the present day one must consult, not a biologist, but certain types of philosopher, or, curiously enough, certain types of mathematical physicist. In histories of biology in the dim future there will probably be a chapter entitled "The Struggle for Existence of the Concept of Organism in the Early Twentieth Century," which will relate how this concept came to be neglected on account of the influence of Descartes, how the metaphysics of natural science in the Nineteenth Century so completely dazzled biologists that they never dreamed of regarding organisms as being anything but swarms of little invisible hard lumps in motion, and how the first blossoming of the concept of organism towards the end of the century was nipped in the bud by the mismanagement of those who advocated it. The rest of the chapter remains to be written. It is of no use dogmatizing or getting emotional about it.

The story is tolerably simple when it is disengaged from the learned verbiage in which it is apt to be concealed. Descartes, who invented the machine theory, understood perfectly well that a machine presupposes a mechanic, but where was the mechanic of the living machine? Descartes was a pious man, and was in the habit of appealing to God to get him out of difficulties. It was to God that he appealed to overcome the difficulty of discriminating between dreams and waking experience, and thus to reassure himself of the existence of an external world, after the celebrated method of doubt had been pushed too far. It was to God that Descartes appealed to furnish the missing mechanic for the organic machines. That was in the days when "development" meant "evolutio." Thus Descartes set the fashion of regarding the organism as a machine with a transcendent mechanic or "organizing principle" and this practice has been followed by some biologists ever since. But others—and they form the majority—who had no use for transcendent principles, and who lacked the Cartesian consistency, have contrived to get on with a machine *without* a mechanic, and this, as Euclid says, is absurd. But nevertheless this view *worked*: that was the main thing, although it did not alter the fact that it was absurd. Only a few thoughtful people have seen that it is absurd. To Driesch belongs the credit of seeing this more clearly than anyone else and of keeping people constantly uneasy about it. But instead of throwing over the machine theory he retains it with a transcendent mechanic, although not in the Cartesian manner. This naturally so disgusted the rank and file that they failed to give sufficient attention to Driesch's admirable critical work. Thus the concept of organism failed to receive due attention because Driesch's mechanic was of no help from the scientific standpoint, however successful it may have been from the metaphysical. But nevertheless some of Driesch's arguments are important and we cannot refute them merely by expressing our dislike of their supposed consequences.

In England the story has taken a different course. Here the shortcomings of the machine theory have been appreciated by a physiologist, J. S. Haldane. He has perceived not only the contradicitions of the machine theory but also the difficulties of appealing to a transcendent mechanic. Accordingly, he has urged the abandonment of the machine theory and the desirability of treating the organism *as* an organism. Why is it then that he has had just as small a following as Driesch? The chief reason seems to be that J. S. Haldane has been content to use the "concept of organism" intuitively in his own work without attempting to give it an abstract formulation. His critical arguments are not so compelling as those of Driesch for a similar reason, and his assertions regarding what he wishes to put in the place of the machine theory are somewhat vague, and require "sympathetic" interpretation. Both of these eminent writers seem to have shown a deplorable lack of understanding of the psychology of the scientific investigator. Moreover, obstacles have also been created by the horrible ambiguity attaching to the term "mechanistic." Abandoning the machine theory does not necessarily mean abandoning "mechnism" in *all* its various meanings, as J. S. Haldane seems to suppose. Neither is "physical" synonymous with "mechanical." It is important to understand this. But Haldane has had the depth of vision to see that far more than a mere re-shuffling or re-definition of biological concepts is involved in the reform of biological thought. It requires investigations which go below the specifically biological level into problems concerning "knowledge about knowledge itself," and it is the difficulties attendant upon an intellectual upheaval of this kind which are largely responsible for the backwardness of the revolution in biological thinking. We do not always appreciate the harm done by the camp followers of great men, in whose hands discoveries which were at first fluid and tentative become petrified into dogmas which create obstacles to further progress. In this way the logical dogmas of Aristotle, the psychological dogmas of Locke, the biological dogmas of Darwin, and above all, the physical and metaphysical dogmas of Descartes are all operative to-day even in the biological sciences. Such dogmas may constitute intellectual blind-spots preventing us from realizing to what a great extent we are free to explore new ways of thinking if new empirical data should require them. But as tentative assumptions harden into dogmas which are never explained our thoughts become encrusted with layers of intellectual rubbish which require the labors of an intellectual Hercules for their proper purgation. Among modern writers who have felt the need for such an undertaking few are better equipped than A. N. Whitehead, and the reader who wishes to learn what it involves, and why it is necessary, cannot do better than read the splendid first two chapters of that author's *Concept of Nature* (14). I have discussed some of these problems from the biological standpoint in a recent publication (18).

Perhaps we can apply G. K. Chesterton's remark about Christianity to the "concept of organism": "It has not been tried and found wanting; it has been found difficult and not tried."

III

After these necessary preliminaries I turn now to the chief task of this paper. I wish first to introduce and define clearly some concepts which appear to be involved in the "concept of organism," to analyse a type of logical order which appears to be exemplified in the organic realm, to examine the use of the causal postulate in biology, and then to apply all this to certain difficulties in embryology and genetics in order to trace them to their sources, and to enable us to think clearly about them. It is hopeless to work in a fog from lack of proper analysis of your thought, and it is equally hopeless to pursue a policy of obscurantism which refuses to allow awkward questions to be raised and openly examined. The situation is complicated and its difficulties are enhanced by the impossibility of saying everything at once. It will be necessary to explain first a number of seemingly disconnected notions and then to try to bring them together at the end

The concept of organism requires a number of subsidiary notions such as "organic whole," "organic part" and "organic relation." Also an organism exhibits what I shall call "hierarchical order." It is easy enough to see "intuitively" what is meant by these terms; there has been a good deal of vague talk from time to time about "the whole being more than the sum of its parts," etc.; the difficulty is to make these notions precise in order to enable us to see how we can use them for scientific purposes. Intuition is the indispensable cutting edge of intellectual inquiry, but the ground won is not consolidated until it has passed from the stage of intuitive apprehension to that of logical analysis.

I shall first try to state quite abstractly what is intended by "hierarchical order," and by abstractly I mean without reference to any *particular* exemplification. We have to investigate hierarchical order as it is in the "logical realm" in its most generalized form. Only then can we profitably discuss its empirical exemplifications. Any reader who finds difficulty in following the abstract exposition can easily make for himself a "particular exemplification" by simply drawing a square on a piece of paper, dividing it into four quarter squares, then dividing each of these into four and so on *ad libitum.* But it must be remembered that this *is* only a particular exemplification, and that hierarchical order has, as such, nothing to do with "space."

The notion of order requires "individuals" or single entities for thought, classes of such, and relations. In hierarchical order we begin with a single individual which will be symbolized by W. This is analysable into individuals called members (m or M) which fall into classes of two kinds called *levels* (L), and *assemblages* (A). There is also a fundamental relation (RiH) in which the members stand to one another, and upon which the whole hierarchical type of order depends. (This will be the relation "being a quarter of" in the case of the square.) We proceed to the following definitions:

Level: A level is a class of members of W which exhaustively divides W and is such that no member of the class stands in the relation R_H to any other member of the class. In any hierarchy there are at least two levels.

Highest level: One level is such that none of its members stand in the relation R_H to any other member, but each stands in the relation R_H to W. This may be called the highest level.

Lowest level: if there is any level which is such that its members are incapable of further analysis this constitutes the lowest level.

Next highest level: if m is any member of a given level L then there is one and only one level (other than L) containing one and only one member M such that m stands in the relation R_H to M. This is the next highest level above L.

Assemblage: if we take any member M belonging to any level except the lowest it will be analysable into a class of members m_1, m_2, m_3, . . . m_n, which is such that it exhausts the member M from which we begin, all its members stand in the relation R_H to M, and none of them stand in the relation R_H to any other member of W. Such a class obtained by analysing any member is called an assemblage.

Thus each member of a given level L is analysable into an assemblage A of members each standing in the relation R_H to A, and since (by definition) none of the members of A stand in the relation R_H to any other member, they constitute, with the members into which the other members of the level L are analysable, a new level, which is the *level next below L.* In this way we can proceed until we reach the lowest level.

It will be found that a given member m cannot be a member of W without standing in the relation R_H to some member M belonging to the level next above its own level unless m belongs to the highest level.

In addition to the generating relation R_H there will also be relations R_L between members of a level, and relations R_A between members of an assemblage.

It is important to note that a given member can always be regarded from three points of view: (1) from the point of view of its membership of a level, i.e., its R_L relations; (2) from the point of view of its entering, with other members of its level, into the constitution of a member of the next highest level, i.e., its R_H and R_A relations; and (3) from the point of view of its analysis into an assemblage of members of the next lowest level in their R_A relations.

This brief account by no means exhausts the subtleties of hierarchical order but will suffice for present requirements. We can turn now to see how it is exemplified in nature. I follow the view of A. N. Whitehead that nature is primarily analysable into spatio-temporal entities (which he calls "events") which pass, and their characters (his "objects") which endure, i.e., can "be again." If we apply hierarchical order to spatio-temporal entities, then W will be a spatio-temporal *whole,* analysable into members which can properly be called *parts.* Moreover hierarchical order may be generated in nature through a temporal process, and this may happen in two ways: (1) by the coming together into R_H relations of originally separate entities which thus become members of a hierarchy constituting a new whole; or, (2) by the division of an original single whole into parts standing in R_H relations. The first case is exemplified in crystallization, the second in the process of development of a metazoan organism. If by a "cell" you mean a spatio-temporal entity which is such that in visual perception it is characterized by an "object" or "pattern" roughly indicated as analysable into a nucleus and cytoplasm, then it is easy to see that in a metazoan animal the parts recognizable as cells constitute a level which is homogeneous with respect to the type of organization of its members. But there may be, and usually are, parts constituting a higher level, namely parts analysable into cells, and these will be cell-assemblages or *cellular parts*. Also each cell is analysable into an assemblage of *cell-parts*, and these are ultimately analysable into the entities with which chemists deal. In cases where it is important to bear in mind that we are speaking of parts it will be desirable to refer to cells as "non-cellular parts" (following the precedent of those who speak of protozoa as non-cellular), since the cell-type of organization characterizes wholes (e.g., germ-cells and protozoa) as well as parts, and from the organic standpoint it is important not to confuse parts and wholes since they have very different properties. Thus we have three levels recognizable in perception: (1) that of which the members are cellular parts; (2) that of which the members are noncellular parts (cells); and (3) that of which the members are cell-parts. Below this there may of course be any number of levels not accessible to perception, and about which we can only assert hypothetical propositions. It is interesting to note that in a metazoan animal there may be spatial parts which lie outside the organic hierarchy, since they do not stand in an R_H relation to other parts, and parts do not stand in such relations to them, for example, the matrix of cartilage, and such parts are always "dead."

So far we have merely noted some obvious points about hierarchical order in the spatial organism. But this is an abstraction resulting from the separation of space from time, and there is no such thing as a living organism in abstraction from time. The living organism is always a fourdimensional entity with temporal as well as spatial extension. The purely spatial three-dimensional organism is the organism considered at a moment, and a moment is an "ideal" to which we approximate by a progressive diminution of the time-dimension in accordance with the method of extensive abstraction, as Whitehead (15) has shown. If now we conceive the organic realm from the four-dimensional standpoint we shall see other ways in which hierarchical order is exemplified. It is generated through the process of reproduction, by which an organic race comes into being. If we consider this four-dimensionally we see that if we begin with a single organism (or pair of organisms) a hierarchy is generated in which the original starting point constitutes the highest level, the F_1 generation constitutes the next level and so on. The relation R_H is the relation "being an immediate descendent of," and the assemblages in each level except the first two will be "families," i.e., the immediate descendents of a given member (or pair) of the level immediately above the one in question. Such a hierarchy would be a genetic hierachy.

Cell-division, as already mentioned, also generates a hierarchy. Considering this from the four-dimensional standpoint we see that starting from a single cell we obtain a hierarchy in which the highest level is this starting point, and each assemblage contains only two members (when division is binary). The R_h relation will be "being an immediate cell-descendant of." The second level will contain one member (an assemblage of two cells), the next two, and so on. If the cells always divide simultaneously those belonging to the same level will always lie in the same "moment." Now there are two possibilities: (1) either the members characterized as cells are single individual organisms as in the protozoa, or (2) they remain in a certain relation to one another so as to constitute not wholes but parts of one organic whole; we then have the "spatial" hierarchy generated, as already mentioned, in the process of development in metazoa.

When a natural entity is divided there are always two possibilities: (1) the two parts may be equal in the sense of having precisely the same properties; or (2) they may be unequal and have different properties. If the process of division in the protozoa is equal in the above sense the individuals in any level will differ from one another only in consequence of the different environmental contingencies they may encounter. In the case of the metazoa, where a single organic whole with its parts in certain determinate hierarchical relations results from division, we again have the two possibilities of equal or unequal division. If the divisions are always equal (in the sense defined above) in this case we could only, it seems, interpret any differences that might later manifest themselves between the cells by supposing that it resulted from differences in their mutual relations in the organic "spatial" hierarchy. When we are considering the metazoa there are two interpenetrating hierarchies involved: (1) that generated by the process of cell-division, call it the "division hierarchy"; and (2) that resulting from the fact that the members of this hierarchy are related by yet another relation in addition to the one which constitues the RH relation in the division hierarchy.

This relation is that complex multiple relation between the parts in virtue of which they *are* parts and not independent entities. This seems to be what people mean by the "organic relation," and is simply the relation of "being a part in an organic spatial hierarchy." A given individual also occupies a determinate position in a genetic hierarchy.

"Cells" can be classified in various ways, depending on the kinds of difference between them. First there is the kind of difference between say, a smooth muscle cell of a monkey, and a smooth muscle cell of a man, and secondly, there is the kind of difference between a smooth muscle cell of a man and a gland cell of a man. The first difference is one which does not "develop" but merely persists. It may be called a genetic difference, and it is this mode of classification of cells which is of interest to genetics. But genetics as an *experimental* science cannot explain how the monkey-cells came to be different from the man-cells, since this was accomplished through a process which occurred in the the remote past. It can only investigate such differences as they are now, not how they came to be. But the second kind of difference is one which develops and is the concern of the embryologist. For there was a time in the history of a given man when he had no smooth muscle cells and no gland cells. From the embryologist's standpoint, therefore, we can classify cells according to the determinate place they occupy in the division hierarchy. If we follow this method we see that the cells belonging to the highest levels are very similar to one another, and we call them embryonic cells, whereas the cells belonging to the lower levels, which are also later and belong to later temporal slices of the individual history, are very different from one another. The cells of these lowest or later levels appear to be divisible into three classes: (1) tissue cells; (2) cells which we may call "persistent embryonic cells," which are concerned in restitutive processes; and (3) germ-cells. Now tissue cells are distinguished from other cells by possessing cell-parts which have *never before been present* in higher or earlier levels of the particular hierarchy in question, although they will usually have been present in previous members of the *genetic* hierarchy. Thus one of the problems of embryology seems to be to explain how these tissue-cell-parts (e.g., myofibrillae, haemoglobin, secretion-granules, etc.,) come to be developed in the cells, how it is that certain parts appear in some cells, and others in other cells, and how it is that they appear *when* they do.

Something must be said at this point about the term "differentiation." Under this term two quite distinct processes appear to have been confused. By *differentiation* I shall mean simply a process through which two (or more) different parts come into being by the separation unequally between them of something previously present in *one* part (or whole). For example, it is by a process of differentiation in this definite but restricted sense that the cells of the animal pole of a frog's egg come to be different from those of the vegetative pole with respect to their yolk content. But there is another process by which parts may come to be different, namely when parts appear which were not previously present as such at all. Of this we have an example in the process already mentioned whereby cell-parts not previously present make their appearance in cells. I shall call such processes processes of *elaboration*, not differentiation.

It does not seem to me to be correct to define the "prospective potency" of a cell in the way Driesch (3) does as "the *possible* fate of a certain cell, i.e., the totality of possible characters of the adult into which this cell may develop." Because, in the first place, as will be seen below, cells do not develop into characters *at all*, and in the second place because only embryonic cells when they undergo their histological elaboration relatively late in development *can* be said to *develop*, and then it is actually only certain cell-parts that develop. Cells of earlier levels of the divison hierarchy do not develop at all, they merely *divide*; that, and that alone, is their "fate."

We have so far considered only two embryological processes: cell-division, or the elaboration of non-cellular parts, and histological elaboration, or the elaboration of cell-parts. That the former is a process of elaboration becomes clear when we recall that the fertilized ovum has no parts characterized as cells, but only cell-parts, and that after cleavage there is a new level in the spatial hierarchy-constituted of cells as parts standing in certain determinate relations to one another. It seems clear from modern genetical studies that the cells belonging to a given race will only be capable of a certain restricted number of cell-part elaborations and it is not at all difficult to understand that an important rôle in this latter process will be played by the hypothetical "genes" required by geneticists, and regarded by them as *nuclear* cell-parts. These of course do not "develop" but only persist, except in so far as mutation occurs. It would be more correct to say that their *properties* persist. If all *nuclear* divisions are equal (in the sense defined) then, in so far as histological or cell-part elaborations depend upon the nucleus, *every* non-cellular part (in a suitable temporal part of the division hierarchy) will theoretically be capable of undergoing any of the histological elaborations of which the race is capable, unless the latter is heterozygous, in which case the cells of a given individual will be capable of only certain of the possible histological elaborations.

But, at least in the higher metazoa, there is another process of elaboration not yet considered, namely the elaboration of *cellular* parts. It will be seen below that it is this developmental process which presents perhaps the greatest difficulties. But before we can proceed further in this direction we must consider the causal postulate a little, and also devote some attention to the different kinds of "properties" and how they may be affected by the kinds of relations met with in organisms. This is of the greatest importance for the interpretation of the data of modern experimental embryology.

IV

The incautious use of the notion of causation, in accordance with the wholly uncritical practice of common sense, has led to some appalling muddles in the discussion of genetical and embryological topics. Some biologists still seem to believe that a "character" can be "solely due to" or "caused by" either a "gene" or "genetic factor" alone, or by an "environmental stimulus" alone, as though "cause" and "effect" were isolated entities standing in a two-termed relation to one another and were completely indifferent to anything else in the world. A very little reflection should suffice to dispel this illusion. Others make some improvement upon this by saying that a given character is partly dependent upon genetic factors, and partly upon environmental factors. But this too cannot be correct if by genetic factors is meant parts of chromosomes, because no organism consists simply of a mass of chromosomes, and you cannot ignore the rest of the organism without futher ado. The notions of "stimulus" and "response" are also often used very loosely in this connexion with unfortunate consequences. Among Schaxel's *Abhandlungen* a useful discussuion of the use of the notion of stimulus has been contributed by P. Jensen (7). The following remarks about the causal postulate are chiefly confined to the requirements of the present paper.

In any causal investigation we always seem to have two "occasions" or "situations" which are *compared.* An occasion contains three principal constituents: (1) an organism, or part of an organism; (2) an environment; and (3) an observer. Or, we may have only one occasion containing two organisms in the same environment. Some simple examples will be helpful. Consider an occasion containing an organism, an observer and some stimulus, i.e , some environmental change. Suppose the organism gives the response r. On a later occasion containing a later temporal part of the same organism the same stimulus is presented. If the same response is observed we say, in accordance with the causal postulate, that the organism has not changed (with respect to the process in question) since the first occasion. If we note a different response r' we say the organism *has* changed, it is a different organism, and it is because of this difference that the response r' differs from r, although this difference may not be observed but is hypothetical. Suppose now we take a number of fertilized eggs from a certain fish and divide them into two batches: one batch (a) is placed in normal sea-water, the other (b) is placed in water containing some abnormal constituent. If the embryos of (b) all exhibit some definite difference (the same in each embryo) from those in batch (a) we say that this difference is "due to" the abnormal constituent of the environment of (b), or, more correctly, it is causally correlated with the difference between the two environments. If we say that the abnormal constituent *caused* the observed abnormality all we are entitled to mean is that it was one element in a complex which is essential to the result observed. Finally, suppose we have two rabbits—one white and one black—born in one litter. Then we say that this difference was correlated either with a difference between the intrauterine conditions of the two embryos, or (which would be considered more probable) with some difference which was present *throughout* the development of both, i.e., a genetic difference.

Thus in all three cases we have comparison. If we observe a *difference* between the organisms concerned we assume either that the observed difference is correlated with some environmental *difference,* or that it is correlated with a previously existing, but perhaps unobserved, *difference* between the organisms concerned. We always believe that if any change occurs this is because something has happened in the environment, or because a change has been going on unobserved in the organism. We can express all this in a simple symbolism:

(1) $\quad D(A, B).C.D(O_A, O_B)$ or $D(E_A, E_B)$
(2) $\quad D(A, B).C.D(O_A, O_B)$ and $D(E_A, E_B)$

These formulae are sufficiently obvious. "D" simply means "the difference between" whatever follows in brackets, and "C" means "is causally correlated with" what follows it. "A" and "B" are the two occasions (or two organisms in one occasion). *"O"* stands for organism and *"E"* for environment. We also require a third formula for cases in which the *relations* are different in the two occasions:

$$(3) \quad D(A, B).C.D(R_A, R_B)$$

Suppose in two occasions I have a lighted candle and a piece of sealing wax. In *"A"* these two constituents are six feet apart, in *"B"* the wax is in the flame. In the former case the wax is solid, in the latter it is melted. The other correlated difference is then the difference between the relations of the two constituents in the two occasions. The same would hold in embryological situations in which so-called "totipotent parts" are transplated to different situations. It may be noted in passing that because we find a second pair it does not follow that this *is* causally correlated with the first.

We must turn now to properties. I shall use the term in a wide sense to include both perceptible characters, e.g., coat color, which may characterize an organism during a considerable period of its history, and also for a specific change which an organism (or part) may exhibit when it enters as a term in a specific causal relation. The most important point which requires explanation from the standpoint of the "concept of organism" is the difference between what will be called "intrinsic" and "relational" properties. In natural science all causal properties are relational properties in the wide sense, but in considering organic wholes or parts it is necessary to distinguish between *viable* relations, and *lethal* relations, i.e., between those in which the organism (or part) lives and those in which it dies. Now by *intrinsic* properties I mean those which an organism or part exhibits in *all viable* relations, and by *relational* properties I mean those properties a *part* exhibits only in certain *organic* relations. And among intrinsic properties it will be necessary to distinguish between *original* and *acquired* intrinsic properties. Thus in a certain relation a part may acquire a property which persists even out of the relation. This would be an acquired property—acquired relationally but becoming intrinsic. This can be illustrated by analogy with "members" of a "social hierarchy." Consider an Englishman born in England. He will have certain intrinsic properties which he exhibits in all viable relations in common with all men, e.g. breathing, eating, etc. But he will also acquire certain relational properties which depend on the "specific hierarchy" to which he belongs. He will, for example, acquire the relational property of speaking English, but if in early childhood he had been "transplanted" to another "social hierarchy," e.g., to Germany, he would have acquired the different relational property of speaking German. He may also take to the sea and acquire the relational property of being steward on board ship. If he is cast upon a desert island he loses this and other relational properties and reverts to a more generalized type in which perhaps he exhibits chiefly his original intrinsic properties. Now precisely the same thing is true of the members of the various levels in the organic hierarchy constituted by the single individual organism. Robinson Crusoe on his island is parallelled by the isolated cell in tissue culture. The important point is that the relational properties of parts depend upon their specific relations in the hierarchy to which they belong.

Now it may be the case that *all* specific histological elaborations are relational properties in the above sense, but that all the cells belonging to a given homozygous race have identical original *intrinsic* properties. But we must pause to consider certain complications. The intrinsic properties of a cell may depend upon both its cytoplasm *and* its nucleus. If two cells, which are assumed to have "equal" nuclei (in the sense defined) behave differently in the same environment, we should say (in accordance with the causal postulate) that they differed intrinsically in their cytoplasm, since their relations are supposed to be the same. But that intrinsic cytoplasmic difference may have been acquired in consequence of relational differences during development, and would therefore be an acquired relational property. But since it now persists in spite of changed relations (since by hypothesis both cells are in the same environment now) we should have to call it an acquired intrinsic property. We seem to have an example of this in the cardiac muscle fibres. These exhibit the elaboration of the rhythmical contractile property in the course of development and this (on the hypothesis under consideration) will be a relational property. But it may persist for years in tissue culture. Hence it is an acquired intrinsic property which is not lost in viable relations even if the part is removed from its place in the individual hierarchy.

Thus it is possible to conceive that throughout the four-dimensional division hierarchy the nuclei are "equal" (i.e. that all nuclear divisions are purely quantitative, except of course the meiotic ones) and that differences between cells (as far as tissue-cells are concerned) are always a consequence of acquired relational properties (which may become intrinsic). Differences between cells may arise either by differentiation or elaboration, and it is in connection with the latter process that relations and relational differences are so important. If we make these assumptions (which seem to have a good deal of empirical evidence in their favor) we shall be able to interpret and state in precise terms many of the curious results which have emerged from transplantation experiments. Some such view as this seems to me to be forced upon us. It will be seen, after a little reflection, that relational properties are of immense importance in an organism, and that far too little attention has been paid to them. It is here that the machine analogy has put us on a wrong scent. There is not discernible in machines a level constituted of parts which are such that the level is homogeneous with respect to the type of organization of those parts, as is the case with the cell-level, and in which the parts can become different in accordance with their relations in the hierarchy. Moreover the *maintenance* as well as the elaboration of specific cell-parts is in some cases dependent on their relations, as is clearly seen in some tissue-culture experiments. In other words the parts in an organic hierarchy are internally related, whereas in a machine these relations are external or non-constitutive relations. How long are we to persist in refusing to look sheer hard facts in the face, merely in the interests of a seventeenth-century analogy which by now may well have outgrown its usefulness? Sooner or later biology will have to take account of them if there is to *be* any theoretical biology, as contrasted with a "medley of *ad hoc* hypotheses."

Another important point came to light in our consideration of hierarchical order in the abstract. It was noted that a given member can be regarded from three points of view, namely from that of (1) its membership of a level, i.e. its R_L relations; (2) its entering into the constitution of a member of the next highest level, i.e., its R_H and R_A relations; and (3) its analysis into an assemblage of members of a lower level in their R_A relations. Now it has been the custom to regard a given organic part almost exclusively from the third point of view, namely from that of its analysis into parts and their properties. But clearly in an organic hierarchy R_H relations will be of equal importance, otherwise its parts would not be different in isolation from what they are in their place in the hierarchy. In other words they would be externally related to one another, like the parts of a machine, which does not seem to be the case. It is for this reason that E. B. Wilson seems to me to be in error when he writes (16), regarding the "organization" of the germ-cells which he says "determines" the "particular course" of development:

> Nevertheless the only available path towards its exploration lies in the mechanistic assumption that somehow the organization of the germ-cell

must be traceable to the physico-chemical properties of its component substances and the specific configuations which they may assume.

How will this be the case if the fertilized ovum is an organic hierarchy with levels *above* the chemical level? We should expect the properties of the members of the chemical levels to depend upon their R_H relations to members of *higher* levels. In a recent lecture in London on the results of micro-injection experiments it was stated that proteins are not present as such in certain living cells. If this is the case it is precisely what we should expect from the organic standpoint. Also it always seems to me to be dangerous to use the word *must* in natural science. That is an expression which may safely be left to fundamentalists and their friends. Let us set them a good example by saying "may" instead of "must" and "probable" instead of "certain," which is all we are usually entitled to say.

Another instance of the neglect of the implications of the concept of organism in the past is furnished by a recent paper by J. Gray (5), in which he discusses the methodological problems involved in the interpretation of growth curves. In his summary he writes:

> Graphic treatment of the data underlying a typical growth curve is liable to produce errors of considerable magnitude, and often tends to confuse the facts. The units which compose a metazoon's body form a very heterogeneous system, in which the rate of growth of one organ is dependent on that of others. It is, therefore, intrinsically improbable that the behavior of such a system should conform to that of a simple chemical system in which the variables are few in number and capable of accurate analysis. The conception of growth as a simple physico-chemical process should not be accepted in the absence of a very rigid and direct proof; at present, it rests on the results of a process of graphic analysis which is often, if not always, of a relatively inaccurate nature.

This clearly represents the first gleams of the dawn of the concept of organism in one consciousness, but a little reflection would have given us even more than this without huge chunks of algebra and pages of graphs. Who but a very learned man would dream of conceiving the growth of a metazoon as a "simple physico-chemical process"? And yet biologists no more ask for a "proof" of this than fundamentalists ask for a proof of the infallibility of the Scriptures. Surely this is a shining example of what C. D. Broad (2) calls a "silly theory" in the sense of one "which may be held at the time one is writing professionally, but which only an inmate of a lunatic asylum would think of carrying into daily life." Now such silly theories are quite indispensable for scientific progress, but it is not only fatal to carry them into daily life (as people frequently do), it is also fatal to take them too seriously in science, lest they harden into dogmas which obscure our intellectual vision, as this one seems to have done. The whole difficulty in relation to these problems rests very largely, I think, on the deep-seated belief that only the ultimate scientific objects into which an organism is believed to be analysable are "really real" (e.g., electrons) and that everything else is "mere appearance" and can be safely neglected. But this is a big subject and here I can only refer the reader to the book by Whitehead (14) already mentioned. This too is a metaphysical legacy from an earlier age.

V

Returning now to the particular problems of genetics and embryology we first have to ask: What does a geneticist mean by a "character?" Consider, for example, a patch of black on some animal's skin: does he mean by the character the very black patch which we see, or does he perhaps mean a vast number of little pigment granules in the skin? If the latter is meant then he is not talking about a character at all but about certain *parts,* and these parts may differ in their properties (characters) from corresponding parts in a related animal. If you try to avoid this by saying that these granules are analysable into sub-parts which differ only in their relations you still do not escape from the antithesis between parts and their characters or properties. Because your ultimate parts will not be *mere* parts, i.e., spatio-temporal entities, but will be distinguished in some way from other spatio-temporal entities by characters or properties of some sort, otherwise the world would consist of a single uniform spatio-temporal entity with no "things" in it at all. Let us say, then, that *correlated* with the black patch which we see there are also, in certain cells of the skin, little pigment granules which differ in their properties from those in the skin-cells of a related animal which has, say, a *yellow* patch correlated with its pigment granules. Let us try to conceive how this is to be interpreted embryologically and genetically. We go back to the fertilized ova from which the two hierarchies were generated by division. In accordance with the chromosome hypothesis we assume that the chromosomes of these two ova differ from one another in some respect, and that with this the difference between the properties of the two kinds of skin-cell granules is correlated. Cleavage now begins, and we assume that the chromosomes are divided equally at each division. This continues until we have a quantity of cells *all* of which, so far as the nucleus is concerned, are capable of elaborating a certain sort (but only *one* sort in *each* whole embryo, i.e., according to the "genetic" classification of its cells) of pigment granule. But towards the end of development, when these granules begin to make their appearance, they do not *appear* in *all* the cells but only in *some* cells. Consequently (if we are to apply the causal postulate) we shall have to appeal to something else to "make the difference" (as we say in everyday life when we are not being "silly"). We seem to have but two alternatives (or a combination of both): Either it is because the *relations* of the skin cells are different from those of the other non-cellular parts (in accordance with our third causal formula); or because the *cytoplasm* is different in these cells. There are facts which would justify the assumption that *if* the difference is to be traceable to cytoplasmic differences these can themselves only be relational (not original), i.e., acquired through relational differences earlier in ontogeny. Thus the hypothetical animal we are considering might be one of those in which, at a suitable stage, it would be possible to transplant cells from, say, the black to the yellow specimen, and they might then develop, not into skin cells but into neural-tube cells with no pigment granules in them at all. As we are trying to consider a purely hypothetical general case we shall make the assumption that the differences between the pigment cells and the remaining cells of the body are *not* dependent upon original cytoplasmic differences but only on relational differences (including differences resulting perhaps from relations of earlier periods). In other words the property of developing pigment granules (as contrasted say with myofibrillae) is *a relational* property. But the property of developing "black" pigment granules in the one case, and "yellow" ones in the other, is clearly an original *intrinsic* property, and distinguishes *all* the cells of the one race from those of the other. It is these properties with which geneticists deal. This enables us to understand how it might happen (as has been shown to happen in some cases) that cells from the "black" specimen from a part which would not ordinarily exhibit skin cells, might be transplated to a "skinny" situation on the "yellow" embryo, and would there behave "ortsgemäss" but would still elaborate not "yellow" but "*black*" granules. Transplanted parts retain their "species specificity." The skinny situation furnishes the particular *organic relations* requisite for the development of *granules* (as contrasted with other possible histological elaborations), but since the particular color correlated with the granules depends upon the intrinsic properties, not on the relational properties (in the sense here used) of the cells, the particular granules in this case will still be "black."

So far so good. We have seen how it is that all the cells may have a complete stock of genes, and how all of the genes *may* be involved in the elaboration of what is necessary for the manifestation of a single character, *provided* we are willing to recognize the difference bewteen intrinsic and relational properties of non-cellular parts, and the existence and importance of internal realtions in an organic hierarchy. Such possibilities were closed to Weismann because he assumed that the properties of a cell (non-cellular part) depend *only* on its nucleus (at least during development) and thus completely ignored the mutual internal relations between parts. He was therefore driven to assume also that "there are material particles in the germ each of which is to be regarded as the primary *Anlage* of one portion of the organism," and that "the chromatin which controls the properties of cells must be different in each kind of cell." (Thus chromatin takes the place of Descartes' God as the "controlling" mechanic.) In conformity with the age in which he lived Weismann also asserted that these assumptions *must* be accepted as correct *for all time to come!* (12). Dogmatism is a deep-seated vice of human nature not confined to theologians (i.e., not a specific relational property!).

But now comes the difficulty. We have all along been taking these specific organic relations in the hierarchy for granted, whereas these relations themselves develop or come into being and this itself constitutes one of the problems of embryology. Development does not consist simply in a process of cell-division followed by one of histological elaboration. Moreover, we started with a black patch, but we have been considering only its blackness and forgetting its shape, size and (perhaps) symmetrical relations to the whole. These are also characters. In considering the blackness of the patch we could concentrate on one process, namely the elaboration of intra-cellular pigment granules with certain properties (i.e., the property of having a *black* color correlated with them for an observer). This might have been the case even although only one cell had been involved. But our patch may involve thousands of cells, and consequently in dealing with its size, shape, and relation to the whole we cannot confine ourselves to what is happening in individual cells. All this means that there is another developmental process to be considered, namely the elaboration of *cellular* parts, of parts, that is to say, which are analysable into cells in certain determinate relations, and which thus constitute parts belonging to a level above that of the cells. It should be remembered that it is upon these parts and their relations (in so far as they differ in different districts in the whole) that we have been driven to place the "responsibility" for the differences in the histological elaborations of the various non-cellular parts which (in their relations) constitute them.

We must go back again to the beginning of development. Conceive an ovum which undergoes a perfectly equal holoblastic cleavage, resulting in a blastula with a perfectly uniform wall consisting of cells all having precisely the same original *intrinsic* properties. This would be an "harmonious equipotential system." Let it undergo gastrulation. Now the causal postulate requires that there should be some difference between the part of the gastrula wall which is invaginated and the rest—either in the organism or in the environment. But by hypothesis it seems that there is no difference in these two parts of the organism, and if we are to rely on the environment then gastrulation would depend (and which particular part is invaginated would depend) on a mere environmental contingency. And this does not seem to be the case. But, it will be said, it is useless to consider such a hypothetical case because no such thing ever happens—the ovum always exhibits some polarity which, in the blastula, is differentiated among the cells of the two poles and it is this which is responsible for the differences in their behavior. Very well then, let there be some such difference, say of size, as seems to be the case in *Amphioxus,* and let this be a sufficient difference to account for gastrulation, as Assheton tried to show (1). This would be a case of differentiation by unequal cytoplasmic division with respect to volume. But this will only help us over one stile and there are very many to cross. Having got our gastrula we now have two cellular parts established: ectoderm and endoderm. We thus have some scope for different relations in the hierarchy. But is this sufficient? What is to "make the difference" (as common-sense folk always say) between right and left halves, dorsal and ventral surfaces? Moreover, we next require differences between the different parts of the archenteric roof to enable us to understand how it is that some become mesodermal pouches, and one becomes notochordal tissue. We can hardly go back to a cytoplasmic "preformation" for these latter differences, because it seems clear from various experiments that they involve relational properties and depend primarily on the blastoporic lip. If we are to assume that some minute part of the egg-cytoplasm is destined to form the organizator how will this enable us to understand the *differences* in the parts of the region it organizes? And even this will not take us very far. As development proceeds more and more cellular parts are elaborated, and if these are all to be referred back to the egg-cytoplasm the latter will become, it seems, intolerably overcrowded. Are all the cellular parts of a man represented in the minute amount of cytoplasm of the human ovum? What, then, shall we say in those cases in which it is possible to chip off pieces of cytoplasm in any direction without preventing the elaboration of cellular parts? How shall we interpret "identical" twins? Does each blastomere possess a complete preformation? It seems clear that a cytoplasmic preformation is as difficult to believe in as Weismann's nuclear one.

Let us then review the possibilities regarding the modes of elaboration of cellular parts as they appear to our present ways of thinking. We seem driven to choose between: (1) a nuclear preformation; (2) a cytoplasmic preformation; (3) environmental differences during the ontogenetic period; (4) appealing to a transcendent "principle"; (5) rejecting the causal postulate. Of these (1) and (3) seem to me to be excluded by modern experimental data as decisively as anything can be excluded by experiment. I can see no possibility of help from (4). This leaves us with (2) and (5), which are not so simple and clear cut as might be supposed at first sight. Even Driesch shrinks from rejecting the causal postulate, at least in the form of the "principle of sufficient reason." And to suggest such a thing in scientific circles is like uttering one of the nine unprintable Anglo-Saxon monosyllables at a polite tea-party. But nevertheless there is still scope for plenty of discussion regarding the use of the causal postulate in embryology. To do this adequately here would carry us far beyond the confines of a single paper. It should, however, be understood that *if* there are equipotential systems, and *if* we are bound by the principle of sufficient reason, then it would absolutely follow that there were some such entities as Driesch's entelechies. I doubt whether the majority of embryologists have fully realized this. But to explain this fully would also take up too much space. All that seems to remain is some kind or degree of "cytoplasmic preformation." Would it be possible to appeal to just enough cytoplasmic preformation to give us our main axes and the primary germ-layers, and could we from thence forwards appeal on a large scale to relational properties? At first the latter would be involved in the blocking out of the main districts during the earlier periods. When these are "set" in accordance with relational properties they behave *"berkunftsgemäss"* on transplanation; hence their morphogenetic properties are now "acquired intrinsic" although *relationally* acquired, as shown by the totally different result of an earlier transplantation. But within each of these main "blocks" there will presumably be sub-parts which undergo elaboration in accordance with relational properties, until finally we reach histological elaboration as already explained. Moreover this "minimum" cytoplasmic preformation to which we are appealing will itself be the outcome of intra-cellular elaboration in accordance with the ovarian relations furnished by the maternal parent. And this enables us to burden the chromosomes with no more than the duty of providing what is needful (whatever that may be) for the various possible histological elaborations of which a given race is capable. This is all that "genes" need to do apart, of course, from persisting and dividing. This possibility, then, has its attractions. But will it square with "identical" twinning and with cutting and transplating experiments? Does it put too much upon relational properties? These are questions requiring further thought and experiment. Experiments should be devised for testing relational properties in greater detail and with greater precision. Also it is very important to know whether there *really are* equipotential systems in the organic realm. A great deal hangs upon this. To sum up: the crucial problems of embryology seem to be those concerning the three principal modes of elaboration—of cellular, non-cellular and of cell-parts. And it seems perfectly clear that it is the *primary* duty of the embryologist to discover how to interpret the developmental process in terms of the intrinsic properties they exhibit in the several hierarchic organic relations which arise in the course of the elaboration of non-cellular parts, and which are accessible to our observation. And this primary objective is quite independent of any theoretical views we may hold regarding the ultimate "reducibility" of biological to other branches of natural knowledge. It is not the business of the biologist to lose his way in a fog of hypothetical imperceptibles. I agree with the physicist who wrote: "I have no doubt whatever that our ultimate aim must be to describe the sensible in terms of the sensible." If Lavoisier, at the foundation of chemistry, had proceeded on the assumption that it was the business of chemistry to reduce itself to physics there would probably never have been any chemistry. The moral for biology is obvious. There is no short cut to biology if it cuts blindly across the observable facts of biological organization. Not to recognize this means tying biology down either to the "verification" of the metaphysics of Herbert Spencer, or, if it is only interpreted methodologically, to an extremely restricted circle of thought devised, in the first instance, on account of its value in a different sphere. This does not apply, of course, to genuine bio-physics and bio-chemistry in their own fields, but to vague speculations in other branches of biological inquiry which merely borrow their terminology and ape their manners.

The "concept of organism" seems to be forcing itself upon our attention from three directions: First, from that of embryology and physiology. The notion of a morphogenetic "field" in Gurwitsch (6) and Weiss (13) is highly symptomatic of this movement and deserves attention. Secondly, there is the *Gestalt-theorie* beginning in psychology and extending to physics (8). Thirdly, there are those philosophers of evolution—Hobhouse, Alexander, Lloyd Morgan, Smuts and Whitehead—all of whom employ the concept of organism in various ways and degrees not exclusively biological. Perhaps this concept will enable us to understand evolution. Has anyone observed a machine that was capable of evolution *without a mechanic?*

List of Literature

(1) ASSHETON, R. Growth in Length. 1916, p. 67 *et seq.*

(2) BROAD, C. D. The Mind and Its Place in Nature. 1925, p. 5.

(3) DRIESCH, H. The Problem of Individuality. 1914, p. 10.

(4) DURKEN, B. Lehrbuch der Experimentalzoologie, 1928, p. 17.

(5) GRAY, J. The kinetics of growth. Brit. Jour. Exp. Bio., 1929. VI, p. 272.

(6) GURWITSCH, A. Versuch einer syntherischen Biologie. 1923.

(7) JENSEN, P. Reiz, Bedingung und Ursache in der Biologie. 1921.

(8) KOHLER, W. Die physischen Gestalten in Ruhe und im stationären Zustand. 1924.

(9) LEWIS, C. I. A Survey of Symbolic Logic. 1918, p. 371.

(10) SCHAXEL, J. Grundzüge der Theorienbildung in der Biologie. 1922.

(11) WARDEN, C. J. The development of modern comparative psychology. QUART. REV. BIO., 1928, III, p. 486.

(12) WERMANN, A. The Germ-Plasm. (Eng. trans.) 1893, p. 4.

(13) WEISS, P. Mosphodynamik. 1926.

(14) WHITEHEAD, A. N. The Concept of Nature. 1926, pp. 1-48.

(15)———. The Principles of Natural Knowledge. 1925, p. 101 *et seq.*

(16) WILSON, E. B. The Cell. 1925, p. 1037.

(17) WOODGER, J. H. Some problems of biological methodology. Proc. Aristotelian Soc., 1929, p. 331.

(18)———. Biological Principles. 1929.

Ernest Nagel

Ernest Nagel was born in Novemesto, Czechoslovakia, in 1901. He came to the United States in 1911 and attended City College in New York City where he received his B.S. in 1923. He was granted the Ph.D. in 1931 from Columbia University. In 1946 he was appointed John Dewey Professor of Philosophy at Columbia University, a chair which he held until 1966 when he became University Professor, and then, in 1976, Professor Emeritus. He is a member of the American Philosophical Society, American Philosophical Association of which he was Eastern Division president in 1954, the Association for Symbolic Logic, American Academy of Arts and Sciences, and a Fellow of the American Association for the Advancement of Science. He is the author of numerous works some of which are: *Sovereign Reason, Logic Without Metaphysics, Gödel's Proof* (with J. R. Newman), and *The Structure of Science*.

The following selection is taken from *The Structure of Science*, (1961), pp. 428-46, and is reprinted here with the kind permission of Harcourt Brace Jovanovich, Inc., New York.

The Standpoint of Organismic Biology

Vitalism of the substantive type advocated by Driesch and other biologists during the preceding century and the earlier decades of the present one is now almost entirely a dead issue in the philosophy of biology. The issue has ceased to be focal, perhaps less as a consequence of the methodological and philosophical criticisms to which vitalism has been subjected, than because of the sterility of vitalism as a guide in biological research and the superior heuristic value of other approaches to the study of vital phenomena. Nevertheless, the historically influential Cartesian conception of biology as simply a chapter of physics continues to meet resistance. Many outstanding biologists who find no merit in vitalism are equally dubious about the validity of the Cartesian program; and they sometimes advance what they believe are conclusive

215

reasons for affirming the irreducibility of biology to physics and the intrinsic autonomy of biological method. The standpoint from which this antivitalistic and yet antimechanistic thesis is currently advanced commonly carries the label of "organismic biology." The label covers a variety of special biological doctrines that are not always mutually compatible. Nonetheless, the doctrines falling under it generally share the common premise that explanations of the "mechanistic" type are not appropriate for vital phenomena. We shall now examine the main contentions of organismic biology.

1. Although organismic biologists deny the suitability if not always the possibility of "mechanistic theories" for vital processes, it is frequently not clear what it is they are protesting against. But such unclarity can undoubtedly be matched by the ambiguity that often marks the statements of aims and programs by professed "mechanists" in biology. As we had occasion to note in an earlier chapter, the word "mechanism" has a variety of meanings, and "mechanists" in biology as well as their opponents take few pains to make explicit the sense in which they employ it. There are biologists who profess themselves to be mechanists simply in the broad sense that they believe that vital phenomena occur in determinate orders and that the conditions for their occurrence are spatiotemporal structures of bodies. But such a view is compatible with the outlook of all schools in biology, with the exception of the vitalists and radical indeterminists; and in any case, when mechanism in biology is so understood, no issue divides those who profess it from most organismic biologists. There have also been biologists who proclaimed themselves to be mechanists in the sense that they maintained that all vital phenomena were explicable exclusively in terms of the science of mechanics (more specifically, in terms of either pure or unitary mechanical theories in the sense of Chapter 7), and who therefore believed living things to be "machines" in the original meaning of this word. It is doubtful, however, whether any biologists today are mechanists in this sense. Physicists themselves have long since abandoned the seventeenth-century hope that a universal science of nature could be developed within the framework of the fundamental ideas of classical mechanics. And it is safe to say that no contemporary biologist subscribes literally to the Cartesian program of reducing biology to the science of mechanics, and especially to the mechanics of contact action.

In any event, most biologists today who call themselves mechanists profess a view that is at once much more specific than the general thesis of causal determinism, and much less restrictive than the one which identifies a mechanistic explanation with an explanation in terms of the science of mechanics. A mechanist in biology, we shall assume, is one who believes, as did Jacques Loeb, that all living processes "can be unequivocally explained in physicochemical terms,"[1] that is, in terms of theories and laws which by common consent are classified as belonging to physics and chemistry. However, biological mechanism so understood must not be taken to deny that living bodies have highly complex organizations. On the contrary, most biologists who adopt such a standpoint usually note quite emphatically that the activities of living bodies are not explicable by analyzing "merely" their physical and chemical compositions without taking into account their "ordered structures or organization." Thus, Loeb's characterization of a living body as a "chemical machine" is an obvious recognition of such organization. It is recognized even more explicitly by E. B. Wilson, who declares, after defining the "development" of germ plasm as the totality of operations by which the germ gives rise to its typical product, that the particular course of this development

> is determined (given the normal conditions) by the specific "organization" of the germ-cells which form its starting-point. As yet we have no adequate conception of this organization, though we know that a very important part of it is represented by the nucleus. . . . Its nature constitutes one of the major unsolved problems of nature. . . . Nevertheless the only available path toward its exploration lies in the mechanistic conception that somehow the organization of the germ-cell must be traceable to the physico-chemical properties of its component substances and the specific configurations which they may assume.[2]

If such is the content of current biological mechanism, and if organismic biologists, like mechanists, reject the postulation of nonmaterial "vitalistic" agents whose operations are to explain vital processes, in what way do the approach and content of organismic biology differ from those of mechanism? The main points of difference, as noted by organismic biologists themselves, appear to be the following:

a. It is a mistake to suppose that the sole alternative to vitalism is mechanism. There are sectors of biological inquiry in which physicochemical explanations play little or no role at present, and a number of biological theories have been successfully exploited which are not physicochemical in character. For example, there is available an impressive body of experimental knowledge concerning embryological processes, though few of the regularities that have been discovered can be explained at present in exclusively physicochemical terms; and neither the theory of evolution even in its current forms, nor the gene theory of heredity, is based on any definite physicochemical assumptions concerning vital processes. It is certainly not inevitable that mechanistic explanations will eventually prevail in these domains; and, since in any event these domains are now being fruitfully explored without any necessary commitment to the mechanistic thesis, organismic biologists possess at least some ground for their doubts concerning the ultimate triumph of that thesis in all sectors of biology. For just as physicists may be warranted in holding that some branch of physics (e.g. electromagnetic theory) is not reducible to some other branch of the science (e.g., to mechanics), so an organismic biologist may be warranted in espousing an analogous view with respect to the relation of biology to the physical sciences. Thus there is a genuine alternative in biology to both vitalism and mechanism—namely, the development of systems of explanation that employ concepts and assert relations neither defined in nor derived from the physical sciences.

b. However, organismic biologists generally claim far more than this. Many of them also maintain that the analytic methods of the physicochemical sciences are intrinsically unsuited to the study of living organisms; that the central problems connected with vital processes require a distinctive mode of approach; and that, since biology is inherently irreducible to the physical sciences, mechanistic explanations must be rejected as the ultimate goal of biological research. One reason commonly advanced for this more radical thesis is the "organic" nature of biological systems. Indeed, perhaps the dominant theme upon which the writings of organismic biologists play so many variations is the "integrated," "holistic," and "unified" character of a living thing and its activities. Living creatures, in contrast to nonliving systems, are not loosely jointed structures of independent and separable parts, are not assemblages of tissues and organs standing in merely external relations to one another. Living creatures are "wholes" and must be studied as "wholes"; they are not mere "sums" of isolable parts, and their activities cannot be understood or explained if they are assumed to be such "sums." But mechanistic explanations construe living organisms as "machines" possessing independent parts, and thereby adopt an "additive" point of view in analyzing vital phenomena. Accordingly, since the action of the whole organism "has a certain unifiedness and completeness" which is left out of account in the course of analyzing it into its elementary processes, E. S. Russell concludes that "the activities of the organism as a whole are to be regarded as of a different order from physico-chemical relations, both in themselves and for the purposes of our understanding."[3] Therefore biology must observe two "cardinal laws of method": "The activity of the whole cannot be fully explained in terms of the activities of the parts isolated by analysis"; and "No part of any living entity and no single process of any complex organic unity can be fully understood in isolation from the structure and activities of the organism as a whole."[4]

c. An additional though closely related point which organismic biologists stress is the "hierarchical organization" of living bodies and processes. Thus, a cell is known to be a structure of various constituents, such as the nucleus, the Golgi bodies, and the membranes, each of which may be analyzable into other parts and these in turn into still others, so that the analysis presumably terminates in molecules, atoms, and their "ultimate" parts. But in multicellular organisms the cell is also only an element in the organization of a tissue, the tissue is a part of some organ, the organ a member of an organ system, and the organ system a constituent in the integrated organism. It is patent that these various "parts" do not occur at the same "level" of organization. In consequence, organismic biologists place great stress on the fact that an animate body is not a system of parts homogeneous in complexity of organization, but that on the contrary the "parts" into which an organism is analyzed must be distinguished according to the different levels of some particular type of hierarchical structure (there may be several such types) to which the parts belong. Now organismic biologists do not deny that physicochemical explanations are possible for the activities of parts on the "lower" levels of a hierarchy. Nor do they deny that the physicochemical properties of the parts on lower levels "condition" or "limit" in various ways the occurrence and modes of action of higher levels of organization. They do deny, on the other hand, that the processes found at higher levels of a hierarchy are "caused" by, or are fully explicable in terms of, lower-level properties. Biochemistry is acknowledged to be the study of the "conditions" under which cells and organisms act the way they do. Organismic biology, on the other hand, investigates the activities of the whole organism "regarded as conditioned by, but irreducible to, the modes of action of lower unities."[5]

We must now examine these alleged differences between the organismic and the mechanistic approaches to biology, and attempt to assess the claim that the mechanistic approach is generally inadequate to biological subject matter.

2. At first blush, the sole issues raised by organismic biology are those we have already discussed in connection with the doctrine of emergence and the reduction of one science to another. In point of fact, other questions are also involved. But to the extent that the issues are those of reduction, we can dispose of them quite rapidly.

Let us first remind ourselves of the two formal conditions, examined at some length in the preceding chapter, that are necessary and sufficient for the reduction of one science to another. When stated with special reference to biology and physiocochemistry, they are as follows:

a. *The condition of connectability.* All terms in a biological law that do not belong to the primary science (such as "cell," "mitosis," or "heredity") must be "connected" with expressions constructed out of the theoretical vocabulary of physics and chemistry (out of terms such as "length," "electric charge," "free energy," and the like). These connections may be of several kinds. The meanings of the biological expressions may be analyzable, and perhaps even explicitly definable, in terms of physicochemical ones, so that in the limiting case the biological expressions are eliminable in favor of the physicochemical terms. An alternative mode of connection is that biological expressions are associated with physicochemical ones by some type of coordinating definition, so that the connections have the logical status of conventions. Finally, and this is the more frequent case, the biological terms may be connected with physicochemical ones on the strength of empirical assumptions, so that the sufficient conditions (and possibly the necessary ones as well) for the occurrence of whatever is designated by the biological terms can be stated by means of the physicochemical expressions. Thus, if the term "chromosome" can be associated in neither of the first two ways with some expression constructed out of the theoretical vocabulary of physics and chemistry, then it must be possible to state in the light of an assumed law the truth-conditions for a sentence of the form "x is a chromosome" entirely by means of a sentence constructed out of that vocabulary.

b. *The condition of derivability.* Every biological law, whether theoretical or experimental, must be logically derivable from a class of statements belonging to physics and chemistry. The premises in these deductions will contain an appropriate selection from the theoretical assumptions of the primary discipline, as well as statements formulating the associations between biological and physicochemical terms required by the condition of connectability. In general, some of the premises will state in the vocabulary of the primary science the boundary conditions or specialized spatiotemporal configuations under which the theoretical assumptions are being applied.

As was shown in the preceding chapter, the condition of derivability cannot be fulfilled unless the condition of connectability is satisfied. It is beyond dispute, however, that the task of satisfying the first of these conditions for biology is still far from completed. We do not know at present, for example, the detailed chemical composition of chromosomes in living cells. We are therefore unable to state in exclusively physicochemical terms the conditions for the occurrence of those organic parts, and hence to state in such terms the truth-conditions for the application of the word "chromosome." And a fortiori we are not able at present to formulate in physicochemical language the structure of any of the systems, such as cell nucleus, cell, or tissue, of which chromosomes are themselves parts. Accordingly, in the current state of biological knowledge it is logically impossible to deduce the totality of biological laws and theories from purely physicochemical assumptions. In short, biology is not at present simply a chapter of physics and chemistry.

Organismic biologists are therefore on firm ground in maintaining that mechanistic explanations of all biological phenomena are currently impossible, and will remain impossible until the descriptive and theoretical terms of biology can be shown to satisfy the first condition for the reduction of that science to physics and chemistry—that is, until the composition of every part or process of living things, and the distribution and arrangement of their parts at any time, can be exhaustively specified in physicochemical terms. Moreover, even if this condition were realized, the triumph of the mechanistic standpoint would not thereby be assured. For as we have already shown, the satisfaction of the condition of connectability is a necessary but in general not a sufficient requirement for the absorption of biology into physics and chemistry. Although the connectability condition might be fulfilled, there would still remain the question whether all biological laws are deducible from the current theoretical assumptions of these physical sciences. The answer to this question is conceivably in the negative, since physicochemical theory in its present form may be insufficiently powerful to permit the derivation of various biological laws, even if these laws were to contain only terms properly linked with expressions belonging to those primary disciplines. It should also be noted that, even if both formal conditions for the reducibility of biology were satisfied, the reduction might nevertheless have little if any scientific importance, for the reason that some of the conditions previously labeled "nonformal" might not be adequately realized.

On the other hand, the facts cited and the argument thus far examined do not warrant the conclusion that biology is *in principle* irreducible to the physical sciences. The task facing such a proposed reduction is admittedly a most difficult one; and it undoubtedly impresses many students as one which, if not utterly hopeless, is at present not worth pursuing. However, no *logical* contradiction has yet been exhibited to the supposition that both the formal and nonformal conditions for the reduction of biology may some day be fulfilled. We can therefore terminate this part of the discussion with the conclusion that the question whether biology is reducible to physicochemistry is an open one, that it cannot be settled by a priori argument, and that an answer to it can be provided only by further experimental and logical inquiry.

3. Let us next turn to the argument for the inherent "autonomy" of biology based on the fact that living systems are hierarchically organized. The burden of the argument, as we have seen, is that properties and modes of behavior occurring on a higher level of such a hierarchy cannot in general be explained as the resultants of properties and behaviors exhibited by isolable parts belonging to lower levels of an organism's structure.

There is no serious dispute among biologists over the thesis that the parts and processes into which living organisms are analyzable can be classified in terms of their respective loci into hierarchies of various types, such as the essentially spatial hierarchy mentioned earlier. Nor is there disagreement over the contention that the parts of an organism belonging to one level of a hierarchy frequently exhibit forms of relatedness and of activity not manifested by organic parts belonging to another level. Thus, a cat can stalk and catch mice, but though the continued beating of its heart is a necessary condition for these activities, the cat's heart cannot perform these feats. Again, the heart can pump blood by contracting and expanding its muscular tissues, although no single tissue can keep the blood in circulation; and no tissue is able to divide by fission, even though its constituent cells may have this property. Such examples suffice to establish the claim that modes of behavior appearing at higher levels of a hierarchically organized system are not explained by merely listing each of the various lower-level parts and processes of the system as an aggregate of isolated and unrelated elements. Organismic biologists do not deny that the occurrence of higher-level traits in hierarchically structured living organisms is contingent upon the occurrence, at different levels of the hierarchy, of various component parts related in definite ways. But they do deny, and with apparent good reason, the statements formulating the traits exhibited by components of an organism, when the components are not parts of an actually living organism, can adequately explain the behaviour of the living system that contains those components as parts related in complex ways to other elements in a hierarchically structured whole.

But do these admitted facts establish the contention that mechanistic explanations are either impossible or unsuitable for biological subject matter? It should be noted that various forms of hierarchical organization are exhibited by the materials of physics and chemistry, and not only by those of biology. Our current theories of matter assume atoms to be structures of electric charges, molecules to be organizations of atoms, solids and liquids to be complex systems of molecules. Moreover, the available evidence indicates that elements at different levels of this hierarchy exhibit traits which their component parts do not invariably possess. However, these facts have not stood in the way of establishing comprehensive theories for the more elementary physical particles and processes, in terms of which it has been possible to account for some if not all of the physicochemical properties exhibited by objects having a more complex organization. To be sure, we do not possess at present a comprehensive and unified theory competent to explain the full range even of purely physicochemical phenomena occurring on various levels of organization. Whether such a theory will ever be achieved is certainly an open question. It is also relevant to note in this connection that biological organisms are "open systems," never in a state of "true equilibrium" but at best only in a steady state of "dynamic equilibrium" with their environment, because they continually exchange material components and not only energies with the latter.[6] In this respect, living organisms are unlike the "closed systems" usually studied in current physical science. Indeed, an adequate theory for physicochemical processes in open systems—for example, a termodynamics competent to deal with systems in nonequilibrium as well as equilibrum states—is at present only in an early stage of development. Nevertheless, the circumstance remains that we can now account for some characteristics of fairly complex systems with the help of theories formulated in terms of relations between relatively more simple ones, for example, the specific heats of solids in terms of quantum theory, or the changes in phase of compounds in terms of the thermodynamics of mixtures. This circumstance must make us pause in accepting the conclusion that the hierarchical organization of living systems by itself precludes a mechanistic explanation for their traits.

Let us, however, examine in greater detail some of the organismic arguments on this issue. One of them has been persuasively stated by J. H. Woodger, whose careful but sympathetic analyses of organismic notions are important contributions to the philosophy of biology. Woodger maintains that it is essential to distinguish between chemical *entities* and chemical *concepts;* he believes that, if the distinction is kept in mind, it no longer appears plausible to assume that a thing can be satisfactorily described in terms of chemical concepts exclusively, merely because the thing is held to be composed of chemical entities. "A lump of iron," Woodger declares, "is a chemical entity, and the word 'iron' stands for a chemical concept. But suppose that the iron has the form of a poker or a padlock, then although the iron is still chemically analyzable in the same way as before it cannot still be fully described in terms of chemical concepts. It now has an organization above the chemical level."[7]

Now there is no doubt that many of the users to which iron pokers or padlocks may be put are not, and may never be, described in purely physicochemical terms. But does the fact that a piece of iron has the form of a poker or of a padlock stand in the way of explaining an extensive class of its properties and modes of behavior in exclusively physicochemical terms? The rigidity, tensile strength, and thermal properties of the poker, or the mechanism and the qualities of endurance of the padlock, are certainly explicable in such terms, even if it may not be necessary or convenient to invoke a microscopic physical theory to account for all these traits. Accordingly, the mere fact that a piece of iron has a certain organization does not preclude the possibility of a physicochemical explanation for some of the characteristics it exhibits as an organized object.

Some organismic biologists maintain that, even if we were able to describe in minute detail the physicochemical composition of a fer-

tilized egg, we would nevertheless still be unable to explain mechanistically the fact that such an egg normally segments. In the view of E. S.
Russell, for example, we might be able on the stated supposition to
formulate the physicochemical conditions for segmentation, but we
would be unable to "explain the course which development takes."[8]

This claim raises some of the previously discussed issues associated
with the distinction between structure and function. But quite apart
from these issues, the claim appears to rest on a misunderstanding if not
on a confusion. It is cogent to maintain that a knowledge of the physicochemical composition of a biological organism does not suffice to
explain mechanistically its modes of action—any more than an enumera-
tion of the parts of a clock together with a description of their spatial
distribution and arrangement suffices to explain or to predict the behavior of the timepiece. To make such an explanation, we must also
assume some theory or set of laws (in the case of the clock, the theory of
mechanics) which formulates the way certain elements act when they
occur in some initial distribution and arrangement, and which permits
the calculation (and hence the prediction) of the subsequent development of that organized system of elements. Moreover, it is conceivable
that, despite our assumed ability at some given stage of scientific
knowledge to describe in full detail the physicochemical composition of
a living thing, we might nevertheless be unable to deduce from the
established physicochemical theories of the day the course of the organism's development. In short, it is conceivable that the first but not the
second formal condition for the reducibility of one science to another is
satisfied at a given time. It is a misunderstanding, however, to suppose
that a fully codified explanation in the natural sciences can consist only
of instantial premises formulating initial and boundary conditions but
containing no statements of law or theory. It is an elementary blunder to
claim that, because some one physicochemical theory (or some class of
such theories) is not competent to explain certain vital phenomena, it is
in principle impossible to construct and establish a mechanistic theory
that can do so.

On the other hand, it would be foolish to underestimate the enormity
of the task facing the mechanistic program in biology because of the
intricate hierarchical organization of living things. Nor should we dismiss as pointless the protests of organismic biologists against versions of
the mechanistic thesis that appear to ignore the fact of such organization. As biologists of all schools have often observed, there is no such
thing as a homogeneous and structurally undifferentiated "living sub-

stance," analogous to "copper substance." There have nevertheless been mechanists who in their statements on biological method, if not in their actual practice as biological investigators, have in effect asserted the contrary. It is therefore worth stressing that the subject matters of their inquiry have compelled biologists to recognize not just a single type of hierarchical organization in living things but several types, and that a central problem in the analysis of organic developmental processes is the discovery of the precise interrelations between such hierarchies.

The hierarchy most frequently cited is generated by the relation of spatial inclusion, as in the case of cell parts, cells, organs, and organisms. However, on any reasonable criterion for distinguishing between various "levels" of such a hierarchy, it turns out that there are bodily parts in most organisms (such as the blood plasma) which cannot be fitted into it. Furthermore, there are types of hierarchy that are not primarily spatial. Thus, there is a "division hierarchy," with cells as elements, which is generated by the division of a zygote and of its cell descendants. Biologists also recognize a "hierarchy of processes": the hierarchy of physicochemical processes in a muscle, the contraction of the muscle, the reaction of a system of muscles, the reaction of the animal organism as a whole; and other types which could be added to this brief list. In any event, it should be noted that in embryological development the spatial hierarchy changes, since in this process new spatial parts are elaborated. This fact can be expressed by saying that, when the division hierarchy of an embryo is compared at different times, its spatial hierarchy at a later time contains elements that did not exist at earlier times. Accordingly, organismic biologists are obviously correct in claiming that to a large extent biological research is concerned with establishing relations of interdependence between various hierarchical structures in living bodies.[9]

Let us now, however, state briefly the schematic form of a hierarchical organization (not necessarily a spatial hierarchy), with a view to assessing in general terms one component in the organismic critique of biological mechanism. Suppose S is some biological system which is analyzable into three major constituents A, B, and C, so that S can be conceived as the relational complex $R(A, B, C)$, where R is some relation. Assume further that each major constituent is in turn analyzable into subordinate constituents $(a_1, a_2, \ldots, a_j)$, $(b_1, b_2, \ldots, b_j)$, and $(c_1, c_2, \ldots, c_k)$, respectively, so that the major constituents of S can be represented as the relational complexes $R(a_1, \ldots, a_j)$, $R(b_1, \ldots, b_j)$, and

$R_c(c_1, \ldots, c_j)$. The a's, b's, and c's may be analyzable still further, but for the sake of simplicity we shall assume only two levels for the hierarchical organization of S. We also stipulate that some of the a's (and similarly some of the b's and c's) stand to each other in various special relations, subject to the condition that all of them are related by R_A to constitute A (with analogous conditions for the b's and c's). Moreover, we assume that some of the a's may stand in certain other special relations to some of the b's and c's, subject to the condition that the complexes A, B, and C are related by R to constitute S. If S is such a hierarchy, one aim of research on S will be to discover its various constituents, and to ascertain the regularities in the relations connecting them with S and with constituents on the same or on different levels.

The pursuit of this aim will in general require the resolution of many serious difficulties. To discover just what the presence of A, for example, contributes to the traits manifested by S taken as a whole, it may be necessary to establish what S would be like in the absence of A, as well as how A behaves when it is not a constitutive part of S. There may be grave experimental problems in attempting to isolate and identify such causal influences. But quite apart from these, the fundamental question must at some point be faced whether the study of A, when it is placed in an environment differing in various ways from the environment provided by S itself, can yield pertinent information about the behaviour of A when it occurs as an actual constituent of S. Suppose, however, that we possess a theory T about the components a of A, such that if the a's are assumed to be in the relation R_A when they occur in an environment E, it is possible to show with the help of T just what traits characterize A in that environment. On this supposition it may not be necessary to experiment upon A in isolation from S. The above crucial question will nevertheless continue to be unresolved unless the theory T permits conclusions to be drawn not only when the a's are in the relation R_A in some artificial environment E, but also when they are in that relation in the particular environment that contains the b's and c's all jointly organized by the relations R_B, R_C, and R. Without such a theory, it will generally be the case that the only way of ascertaining just what role A plays in S is to study A as an actual component in the relational complex $R(A, B, C)$.

Accordingly, organismic biologists are right in insisting on the general principle that "an entity having the hierarchical type of organization such as we find in the organism requires investigation at all levels, and investigation of one level cannot replace the necessity for investigation of levels higher up in the hierarchy."[10] On the other hand, this principle does not entail the impossibility of mechanistic explanations for vital phenomena, though organismic biologists sometimes appear to believe that it does. In particular, if the a's, b's, and c's in the above schematism are the submicroscopic entities of physics and chemistry, S is a biological organism, and T is a physicochemical theory, it is not impossible that the conditions for the occurrence of the relational complexes A, B, C, and S can be specified in terms of the fundamental concepts of T, and that furthermore the laws concerning the behaviors of A, B, C, and S can be deduced from T. But, as has been argued in the preceding chapter, whether in point of fact one science (such as biology) is reducible to some primary science (such as physicochemistry), is contingent on the character of the particular theory employed in the primary discipline at the time the question is put.

4. We must finally turn to what appears to be the main reason for the negative attitude of organismic biologists toward mechanistic explanations of vital phenomena, namely, the alleged "organic unity" of living things and the consequent impossibility of analyzing biological wholes as "sums" of independent parts. Whether there is merit in this reason obviously depends on what senses are attached to the crucial expressions "organic unity" and "sum." Organismic biologists have done little to clarify the meanings of these terms, but at least a partial clarification has been attempted in the preceding and present chapters of this book. In the light of these earlier discussions the issue now under examination can be disposed of with relative brevity.

Let us assume, as do organismic biologists, that a living thing possesses an "organic unity," in the sense that it is a teleological system exhibiting a hierarchical organization of parts and processes, so that the various parts stand to each other in complex relations of causal interdependence. Suppose also that the particles and processes of physics and chemistry constitute the elements at the lowest level of this hierarchical system, and that T is the current body of physicochemical theory. Finally, let us associate with the word "sum" in the statement "A living organism is not the sum of its physicochemical parts," the "reducibility" sense of the word distinguished in the preceding chapter. The statement will then be understood to assert that, even when suitable physicochemical initial and boundary conditions are supplied, it is not possible to deduce from T the class of laws and other statements about living things commonly regarded as belonging to the province of biology.

Subject to an important reservation, the statement construed in this manner may very well be true, and probably represents the opinion of most students of vital phenomena, whether or not they are organismic biologists. The statement is widely held, despite the fact that in many cases physicochemical conditions for biological processes have been ascertained. Thus, an unfertilized egg of the sea urchin does not normally develop into an embryo. However, experiments have shown that, if such eggs are first placed for about two minutes in sea water to which a certain quantity of acetic acid has been added and are then transferred to normal sea water, the eggs presently begin to segment and to develop into larvae. But, although this fact certainly counts as impressive evidence for the physicochemical character of biological processes, the fact has thus far not been fully explained, in the strict sense of "explain," in physicochemical terms. For no one has yet shown that the statement that sea urchin eggs are capable of artificial parthenogenesis under the indicated conditions is *deducible* from the purely physicochemical assumptions *T*. Accordingly, if organismic biologists are making only the *de facto* claim that no systems possessing the organic unity of living things have thus far been proved to be sums (in the reducibility sense) of their physicochemical constituents, the claim is undoubtedly well founded.

On the other hand, in the prevailing circumstances of our knowledge there should be no cause for surprise that the fact about the artificial parthenogenesis of sea urchin eggs is not deducible from *T*. The deduction is not possible, if only because the elementary logical requirements for performing it are currently not satisfied. No theory can explain the operations of any concretely given system unless a complete set of initial and boundary conditions for the application of the theory is stated in a manner consonant with the specific notions employed in the theory. For example, it is not possible to deduce the distribution of electric charges on a given insulated conductor merely from the fundamental equations of electrostatic theory. Additional instantial information must be supplied in a form prescribed by the character of the theory—in this instance, information about the shape and size of the conductor, the magnitudes and distribution of electric charges in the neighborhood of the conductor, and the value of the dielectric constant of the medium in which the conductor is embedded. In the case of the sea urchin eggs, however, although the physicochemical composition of the environment in which the unfertilized eggs develop into embryos is presumably known, the physicochemical composition of the eggs themselves is still unknown, and cannot be formulated for inclusion in the indispensable instantial conditions for the application of T. More generally, we do not know at present the detailed physicochemical composition of any living organism, nor the forces that may be acting between the elements on the lowest level of its hierarchical organization. We are therefore currently unable to state in exclusively physicochemical terms the initial and boundary conditions, requisite for the application of T to vital systems. Until we can do this, we are in principle precluded from deducing biological laws from mechanistic theory. Accordingly, although it may indeed be true that a living organism is not the sum of its physicochemical parts, the available evidence does not warrant the assertion either of the truth or of the falsity of this dictum.

Although the point just stressed is elementary, organismic biologists often appear to neglect it. They sometimes argue that, while mechanistic explanations may be possible for traits of organic parts when these parts are studied in "abstraction" (or isolation) from the organism as a whole, such explanations are not possible when the parts function conjointly in mutual dependence as actual constituents of a living thing. But this claim ignores the crucial fact that the initial conditions required for a mechanistic explanation of the traits of organic parts manifested when the parts exist *in vitro* are generally insufficient for accounting mechanistically for the conjoint functioning of the parts in a biological organism. For it is evident that when a part is isolated from the rest of the organism it is placed in an environment which is usually different from its normal environment, where it stands in relations of mutual dependence with other parts of the organism. It therefore follows that the initial conditions for using a given theory to explain the behavior of a part in isolation will also be different from the initial conditions for using that theory to explain behavior in the normal environment. Accordingly, although it may indeed be beyond our actual competence at present or in the foreseeable future to specify the instantial conditions requisite for a mechanistic explanation of the functioning of organic parts *in situ,* there is nothing in the logic of the situation that limits such explanations in principle to the behavior of organic parts *in vitro.*

One final comment must be added. It is important to distinguish the question of whether mechanistic explanations of vital phenomena are possible from the quite different though related problem of whether living organisms can be effectively synthesized in a laboratory out of nonliving materials. Many biologists seem to deny the first possibility because of their skepticism concerning the second. In point of fact, however, the two issues are logically independent. In particular, although it may never become possible to manufacture living organisms artificially, it does not follow that vital phenomena are therefore incapable of being explained mechanistically. A glance at the achievements of the physical sciences will suffice to establish this claim. We do not possess the power to manufacture nebulae or solar systems, despite the fact that we do possess physicochemical theories in terms of which nebulae and planetary systems are tolerably well understood. Moreover, while modern physics and chemistry provide competent explanations for various properties of chemical elements in terms of the electronic structure of the atoms, there are no compelling reasons for believing that, for example, men will some day be capable of manufacturing hydrogen by putting together artificially the subatomic components of the substance. On the other hand, the human race possessed skills (e.g., in the construction of dwellings, in the manufacture of alloys, and in the preparation of foods) long before adequate explanations for the traits of the artifically constructed articles were available.

Nonetheless, organismic biologists often develop their critique of the mechanistic program in biology as if its realization were equivalent to the acquisition of techniques for literally taking apart living things and then overtly reconstituting the original organisms out of their dismembered and independent parts. However, the conditions for achieving mechanistic explanations for vital phenomena are quite different from the requirements for the artificial manufacture of living organisms. The former task is contingent on the construction of factually warranted theories of physicochemical substances; the latter task depends on the availability of suitable physicochemical materials, and on the invention of effective techniques for combining and controlling them. It is perhaps unlikely that living organisms will ever be synthesized in the laboratory except with the help of mechanistic theories of vital processes; in the absence of such theories, the artifical manufacture of living things, were this ever accomplished, would be the outcome of a fortunate but improbable accident. But in any event, the conditions for achieving these patently different tasks are not identical, and either may some day be realized without the other. Accordingly, a denial of the possibility of mechanistic explanations in biology on the tacit supposition that these conditions do coincide, is not a cogently reasoned thesis.

*   *   *

The main conclusion of this discussion is that organismic biologists have not established the absolute autonomy of biology or the inherent impossibility of physicochemical explanations of vital phenomena. Nevertheless, the stress they place on the hierarchical organization of living things and on the mutual dependence of organic parts is not a misplaced one. For, although organismic biology has not convincingly secured all its claims, it has demonstrated the important point that the pursuit of mechanistic explanations for vital processes is not a *sine qua non* for valuable and fruitful study of such processes. There is no more reason for rejecting a biological theory (e.g., the gene theory of heredity) because it is not a mechanistic one (in the sense of "mechanistic" we have been employing) than there is for discarding some physical theory (e.g., modern quantum theory) on the ground that it is not reducible to a theory in another branch of physical science (e.g. to classical mechanics). A wise strategy of research may indeed require that a given discipline be cultivated as a relatively independent branch of science, at least during a certain period of its development, rather than as an appendage to some other discipline, even if the theories of the latter are more inclusive and better established than are the explanatory principles of the former. The protest of organismic biology against the dogmatism often associated with the mechanistic standpoint in biology is salutary.

There is, however, an obverse side to the organismic critique of that dogmatism. Organismic biologists sometimes write as if any analysis of vital processes into the operation of distinguishable parts of living things entails a seriously distorted view of these processes. For example, E. S. Russell has maintained that in analyzing the activities of an organism into elementary processes "something is lost, for the action of the whole has a certain unifiedness and completeness which is left out of account in the process of analysis."[11] Analogously, J. S. Haldane claimed that we cannot apply mathematical reasoning to vital processes, since a mathematical treatment assumes a separability of events in space "which does not exist for life as such. We are dealing with an indivisible whole when we are dealing with life."[12] And H. Wildon Carr, a professional philosopher who subscribed to the organismic standpoint and wrote as one of its exponents, declared that "Life is individual; it exists only in living beings, and each living being is indivisible, a whole not constituted of parts."[13]

Such pronouncements exhibit an intellectual temper that is as much an obstacle to the advancement of biological inquiry as is the dogmatism of intransigent mechanists. In biology as in other branches of science knowledge is acquired only by analysis or the use of the so-called "abstractive method"—by concentrating on a limited set of properties things possess and ignoring (at least for a time) others, and by investigating the traits selected for study under controlled conditions. Organismic biologists also proceed in this way, despite what they may say, for there is no effective alternative to it. For example, although J. S. Haldane formally proclaimed the "indivisible unity" of living things, his studies on respiration and the chemistry of the blood were not conducted by considering the body as an indivisible whole. His researches involved the examination of relations between the behavior of one part of the body (e.g. the quantity of carbon dioxide taken in by the lungs) and the behavior of another part (the chemical action of the red blood cells). Like everyone else who contributes to the advance of knowledge, organismic biologists must be abstractive and analytical in their research procedures. They must study the operations of various prescinded parts of living organisms under selected and often artificially instituted conditions—on pain of mistaking unenlightening statements liberally studded with locutions like "wholeness," "unifiedness," and "indivisible unity" for expressions of genuine knowledge.

NOTES

[1]Jacques Loeb, *The Mechanistic Conception of Life,* Chicago, 1912.

[2]E. B. Wilson, *op. cit.,* p. 1037, quoted by kind permission of The Macmillan Company, New York.

[3]E. S. Russell, *The Interpretation of Development and Heredity,* Oxford, 1930, pp. 171-72.

[4]*Ibid.,* pp. 146-47. Similar statements of the central tenet of organismic biology will be found in Russell's *Directiveness of Organic Activities,* Cambridge, England, 1945, esp. Chaps. 1 and 7; Ludwig von Bertalanffy, *Theoretische Biologie,* Berlin, 1932, Chap. 2; his *Modern Theories of Development,* Oxford, 1933, Chap. 2; and his *Problems of Life,* New York and London, 1952, Chaps. 1 and 2; and W. E. Agar, *The Theory of the Living Organism,* Melbourne and London, 1943.

[5]Russell, *The Interpretation of Development and Heredity,* p. 187. For an analogous view, cf. Ludwig von Bertalanffy and Alex B. Novikoff, "The Conception of Integrative Levels and Biology," *Science,* Vol. 101 (1945), pp. 209-15, and the discussion of this article in the same volume, pp. 582-85 and in Vol. 102 (1945), pp. 405-06. A careful and sober analysis of the nature of hierarchical organization in biology and of its import for the possibility of mechanistic explanation is given in J. H. Woodger, *Biological Principles,* New York, 1929, Chap. 6 and in Woodger's "The 'Concept of Organism' and the Relation between Embryology and Genetics," *Quarterly Review of Biology,* Vol. 5 (1930), and Vol. 6 (1931).

[6]L. von Bertalanffy, *Problems of Life*, Chap. 4.

[7]J. H. Woodger, *Biological Principles*, p. 263. Woodger continues, "In the same way an organism is a physical entity in the sense that it is one of the things we become aware of by means of the senses, and is a chemical entity in the sense that it is capable of chemical analysis just as is the case with any other physical entity, but it does not follow from this that it can be fully and satisfactorily described in chemical terms."

[8]E. S. Russell, *The Interpretation of Development and Heredity*, p. 186.

[9]Cf. the writings of Woodger cited above, as well as his *Axiomatic Method in Biology*, Cambridge, Eng., 1937; also L. von Bertalanffy, *Problems of Life*, Chap. 2.

[10]J. H. Woodger, *Biological Principles*, p. 316.

[11]E. S. Russell, *The Interpretation of Development and Heredity*, p. 171.

[12]J. S. Haldane, *The Philosophical Basis of Biology*, London, 1931, p. 14.

[13]Quoted in L. Hogben, *The Nature of Living Matter*, London, 1930, p. 226.

BIBLIOGRAPHY

Ackermann, Robert. "Mechanism and the Philosophy of Biology," *Southern Journal of Philosophy,* 6 (1968), 143-51.

———. "Mechanism, Methodology, and the Biological Theory," *Synthèse,* 20 (1969), 219-29.

Ayala, Francisco Jose and Theodosius Dobzhansky (eds.). *Studies in the Philosophy of Biology: Reduction and Related Problems.* London: Macmillan, 1974.

Baublys, Kenneth K. "Discussion: Comments on Some Recent Analyses of Functional Statements in Biology," *Philosophy of Science,* 42 (1975), 469-86.

Beauregard, O Costa de. "On Time, Information, and Life," *Dialectica,* 22 (1968), 187-205.

Beckner, Morton. "Aspects of Explanation in Biological Theory." Voice of America Forum Lectures, Philosophy of Science Series No. 13, United States Information Agency, Washington, D.C., 1965.

The Biological Way of Thought. Berkeley: University of California Press, 1968.

———. "Metaphysical Presuppositions and the Descriptions of Biological Systems," *Synthese,* 15 (1963), 260-74.

Bertalanffy, Ludwig von. *Modern Theories of Development,* trans. J. H. Woodger. New York: Harper & Row, 1962.

———. *Perspectives on General Systems Theory: Scientific-Philosophical Studies.* Ed. Edgar Taschdjian. New York: George Braziller, 1975.

———. *Problems of Life.* New York: John Wiley & Sons, 1952.

Blackburn, Robert T. (ed.). *Interrelations: The Biological and Physical Sciences.* Chicago: Scott, Foresman and Company, 1966.

Blandino, Giovanni. *Theories on the Nature of Life.* New York: Philosophical Library, 1969.

Blum, H. F. *Time's Arrow and Evolution.* Princeton: Princeton University Press, 1951.

Bohr, Niels. "Light and Life," *Nature,* 131 (1933), 421-23 and 457-59.

Bonner, J. T. *The Ideas of Biology.* London: Methuen, 1965.

Braithwaite, R. B. *Scientific Explanation.* Cambridge: Cambridge University Press, 1953.

Canfield, John. "Teleological Expalantion in Biology," *British Journal for the Philosophy of Science,* XIV (1963-4), 285-95.

Canfield, John (ed.). *Purpose in Nature.* Englewood Cliffs, N.J.: Prentice-Hall, Inc., 1966.

Capek, Milic. *Bergson and Modern Physics: A Re-interpretation and a Reevaluation.* Dordrecht: Reidel, 1971.

Carlo, William E. "Mechanism and Vitalism, a Reappraisal," *Pacific Philosophical Forum,* 6 (1968), 57-68.

———. "Reductionism and Emergence, Mechanism and Vitalism Revisited," Proceedings of the Catholic Philosophy Association, 40 (1966), 94-103.

Causey, Robert L. "Polanyi on Structure and Reduction," *Synthèse,* 20 (1969), 230-37.

Dobzhansky, Theodosius. *The Biology of Ultimate Concern.* New York: New American Library, 1967.

Dunn, L. C. *A Short History of Genetics.* New York: McGraw-Hill, 1965.

Edelstein, Barry. "Thermodynamics, Kinetics, and Biology," *Zygon,* 6 (1971), 160-62.

Elsasser, Walter M. *Atom and Organism: A New Approach to Theoretical Biology.* Princeton: Princeton University Press, 1966.

———. *The Chief Abstractions of Biology.* Amsterdam: North-Holland Publishing Company, 1975.

———. *The Physical Foundations of Biology: An Analytical Study.* New York: Pergamon Press, 1958.

Feyerabend, Paul. "Explanation, Reduction, and Empiricism," in H. Feigl and G. Maxwell (eds.), *Minnesota Studies in the Philosophy of Science,* Vol. 3. Minneapolis: University of Minnesota Press, 1962, pp. 28-97.

Frankfurt, Harry G. and Brian Poole. "Functional Analyses in Biology," *British Journal for the Philosophy of Science,* 17 (1966), 69-72.

Frolov, I. T. "The Nature of Contemporary Biological Knowledge: Methodological Analysis," *Soviet Studies in Philosophy,* 12 (1973-74) 27-48.

Geller, E. S. and E. G. Babskii. "Cybernetics and Life," *Soviet Studies in Philosophy,* 8 (1970), Gierer, D. A. "The Physical Foundations of Biology and the Problems of Psychophysics," *Ratio,* 12 (1970), 47-64.

Grene, Marjorie. *Approaches to a Philosophical Biology.* New York: Basic Books, 1968.

———. "Aristotle and Modern Biology," *Journal of the History of Ideas,* 33 (1972), 395-424.

———. *The Understanding of Nature: Essays in the Philosophy of Biology.* Boston Studies in the Philosophy of Science, Vol. XXIII, Boston: Reidel, 1974.

Grene, Marjorie, and Everett Mendelsohn (eds.). *Topics in the Philosophy of Biology.* Boston Studies in the Philosophy of Science, Vol. XXVII, Boston: Reidel, 1976.

Gunter, Pete Addison. "Biological Time and Biological Mechanism: Reflections on the "New Embryology'," *Southwestern Journal of Philosophy*, 2 (1971), 173-83.

———. "The Heuristic Force of Creative Evolution," *Southwestern Journal of Philosophy*, 1 (1970), 111-18.

Haldane, J. S. *Mechanism, Life, and Personality* (2nd ed.). London: John Murray, 1921.

———. *The Philosophical Basis of Biology*. New York: Doubleday, Doran and Company, 1931.

Haraway, Donna Jeanne. *Crystals, Fabrics, and Fields: Metaphors of Organicism in Twentieth Century Developmental Biology*. New Haven: Yale University Press, 1976.

Hein, Hilde. "Mechanism, Vitalism, and Biopocsis," *Pacific Philosophical Forum*, 6 (1968), 4-56.

———. "Molecular Biology vs. Organicism: The Enduring Dispute Between Mechanism and Vitalism," *Synthèse*, 20 (1969), 238-53.

———. *On the Nature and Origin of Life*. New York: McGraw-Hill, 1971.

Hempel, Carl. *Aspects of Scientific Explanation*. New York: The Free Press, 1965.

Hintikka, Jaakko (ed.), *Synthèse* (Vol. 20, No. 2, devoted to philosophy of biology), August 1969.

Hirschman, David. "Function and Explanation, Part I." Proceedings of the Aristotelian Society, 47 (1953), 19-37.

Hull, David L. "What the Philosophy of Biology Is Not," *Synthèse*, 20 (1969), 157-84.

Hull, David L. (ed.). *Philosophy of Biological Science*. Englewood Cliffs, N.J.: Prentice-Hall, 1974.

Kant, Immanuel. *Critique of Teleological Judgement*, trans. J. C. Meredith. Oxford: The Clarendon Press, 1928.

Kleiner, Scott A. "Essay Review: The Philosophy of Biology," *Southern Journal of Philosophy*, 13 (1975), 523-42.

Kochanski, Zdzislaw. "Conditions and Limitations of Prediction Making in Biology," *Philosophy of Science*, 40 (1973), 29-51.

Koestler, Arthur and J. R. Smythies (eds.). *Beyond Reductionism: New Perspectives in the Life Sciences*. Alpbach Symposium, 1968. (1st American ed.), New York: Macmillan, 1970.

Lagerspetz, Kari. "Individuality and Creativity: Is Biology Different?" *Synthèse*, 20 (1969), 254-60.

Lehman, Hugh. "Are Biological Species Real?" *Philosophy of Science*, 34 (1965), 157-67.

———. "Functional Explanation in Biology," *Philosophy of Science*, 32 (1965), 1-20.

———. "On the Form of Expalantion in Evolutionary Biology," *Theoria*, 32 (1966), 14-24.

————. "Teleological Explanations in Biology," *British Journal for the Philosophy of Science*, XV (1964-65), 327.

Loeb, Jacques. *The Mechanistic Conception of Life*. Ed. Donald Fleming. Cambridge: The Belknap Press of Harvard University Press, 1964.

Macklin, Martin, and Ruth Macklin. "Theoretical Biology: A Statement and a Defense," *Synthèse*, 20 (1969), 261-76.

Manier, Edward. "The Experimental Method in Biology," *Synthèse*, 20 (1969), 185-205.

————. " 'Fitness' and Some Explanatory Patterns in Biology," *Synthèse*, 20 (1969), 206-18.

Mayr, Ernst. "Footnotes on the Philosophy of Biology," *Philosophy of Science*, 36 (1969), 197-202.

Mendelsohn, Everett (ed.). *Journal of the History of Biology* (Vol. 2, No. 1, devoted to history and philosophy of biology), 1969.

Moulyn, Adrian C. *Structure, Function, and Purpose: An Inquiry into the Concepts and Methods of Biology from the Viewpoint of Time*. New York: Liberal Arts Press, 1957.

Munson, Ronald. "Is Biology a Provincial Science?" *Philosophy of Science*, 42 (1975), 428-47.

Nagel, Ernest. *The Structure of Science*. London: Routledge and Kegan Paul. 1971.

Needham, Joseph. *Order and Life*. New Haven: Yale University Press, 1936.

Oparin, A. I. *Life: Its Nature, Origin, and Development*, trans. Ann Synge. New York: Academic Press, 1961.

————. "Modern Aspects of the Problem of the Origin of Life," *Scientia*, 65 (1971), 195-206.

Pantin, C. F. A. *The Relations Between the Sciences*. London: Cambridge University Press, 1968.

Partashnikov, Anatoly. "Soviet Philosophy of Biology Today," *Studies in Soviet Thought*, 14 (1974), 1-25.

Phillips, D. C. "Organicism in the Late Nineteenth and Early Twentieth Centuries," *Journal of the History of Ideas*, 31 (1970), 413-32.

Plamondon, Ann. "The Contemporary Reconciliation of Mechanism and Organicism," *Dialectica*, 29 (1975), 213-21.

Polanyi, Michael. "Life's Irreducible Structure," *Science*, 160 (1968), 1308-12.

————. *Personal Knowledge*. New York: Harper & Row, 1964.

Raven, Chri P. "Formalization of the Fundamental Concepts in Some Fields of Biology," *Synthèse*, 7 (1948), 93-99.

————. "Irrational Elements in Some Theories of Life," *Synthèse*, 10 (1952), 359-63.

Robinson, Andrew. "Is There Purpose in Modern Biology?" *Proceedings of the Catholic Philosophy Association*, 46 (1972).

Roll-Hansen, Nils. "On the Reduction of Biology to Physical Science," *Synthèse*, 20 (1969), 277-89.

Ruse, Michael. "Are There Laws in Biology?" *Australasian Journal of Philosophy*, 48 (1970), 234-46.

———. "Functional Statements in Biology," *Philosophy of Science*, 38 (1971), 87-95.

———. *Philosophy of Biology*. London: Hutchinson's University Library, 1973.

Schaffner, Kenneth F. "Antireductionism and Molecular Biology," *Science*, 157 (1967), 644-47.

———. "Approaches to Reduction," *Philosophy of Science*, 34 (1967), 137-47.

———. "Theories and Explanations in Biology," *Journal of the History of Biology*, 2 (1969), 19-33.

Schrödinger, Erwin. *What is Life? The Physical Aspect of the Living Cell, Mind and Matter*. Cambridge: Cambridge University Press, 1967.

Scriven, Michael. "Explanation in the Biological Sciences," *Journal of the History of Biology*, 2 (1969), 187-98.

Shapere, Dudley. "Biology and the Unity of Science," *Journal of the History of Biology*, 2 (1969), 3-18.

Shelanski, V. B. "Nagel's Translation of Teleological Statements: A Critique." *British Journal for the Philosophy of Science*, 24 (1973), 397-401.

Simon, Michael A. *The Matter of Life*. New Haven: Yale University Press, 1971.

Sinnott, Edmund W. *The Biology of the Spirit*. New York: Viking Press, 1955.

———. *The Bridge of Life: From Matter to Spirit*. New York: Simon and Schuster, 1966.

———. *Cell and Psyche: The Biology of Purpose*. Chapel Hill: University of North Carolina Press, 1966.

———. *Matter, Mind, and Man: The Biology of Human Nature*. New York: Atheneum, 1962.

Smith, Christopher. *The Problem of Life: An Essay in the Origins of Biological Thought*. New York: John T. Wiley, 1976.

Smith, Vincent E. *Philosophical Problems in Biology*. New York: St. John's University Press, 1966.

Taylor, Charles. *The Explanation of Behavior*. London: Routledge & Kegan Paul, 1964.

Teilhard de Chardin, Pierre. *The Phenomenon of Man*, trans. Bernard Wall. London: Wm. Collins Sons & Company, 1959.

Varela, Francisco G., and Humberto Maturana. "Mechanism and Biological Explanation," *Philosophy of Science*, 39 (1972), 378-82.

Waddington, C. H. *The Nature of Life*. New York: Harper & Row, 1966.

Waddington, C. H. (ed.). *Towards a Theoretical Biology*. Chicago: Aldine Publishing Comapny, 1968.

Wagener, James W. "From Instinct to Thought: Chardin's Evolutionary Theory of Knowledge," *Journal of Thought*, 5 (1970), 18-20.

Watson, James D. *The Double Helix*. New York: Atheneum, 1968.

————. *The Molecular Biology of the Gene.* New York: W. A. Benjamin, Inc.,

Wimsatt, William C. "Teleology and the Logical Structure of Function Statements," *Studies in the History and Philosophy of Science,* 3 (1972), 1-80.

Wolvekamp, H. "The Concept of the Organism as an Integrated Whole," *Dialectica,* 20 (1966), 196-214.

Woodger, J. H. *The Axiomatic Method in Biology.* Cambridge: Cambridge University Press, 1937.

————. *Biological Principles.* (Reissued with a New Introduction.) London: Routledge & Kegan Paul, 1967.